YOUR GUIDE TO EMERGENCY HOME STORAGE

ALAN K. BRISCOE

First Printing: April 1999
International Standard Book Number
0-88290-660-7

Horizon Publishers' Catalog and Order Number
1223

Printed and distributed
in the United States of America by

Horizon Publishers
& Distributors, Incorporated

Mailing Address:
P.O. Box 490
Bountiful, Utah 84011-0490

Street Address:
50 South 500 West
Bountiful, Utah 84010

Telephone:
Local Phone: (801) 295-9451
WATS (toll free): 1 (800) 453-0812
FAX: (801) 295-0196

E-mail:
horizonp@burgoyne.com

Internet Home Page:
http://www.horizonpublishers.com

CONTENTS

INTRODUCTION

Why Emergency Home Storage?

In this complex age, unexpected events can suddenly disrupt the lives and well-being of families, communities, and even nations—placing people in danger and at risk for extended periods of time. In such circumstances, it is prudent that every family should have an emergency readiness plan.

There are many reasons why you should store food, clothing and other necessities. Among them are:

- ***natural disasters*** such as hurricanes, earthquakes, floods, volcanic eruptions, landslides, epidemics, droughts, and famines,
- ***political situations*** such as wars, riots, terrorist attacks, sabotage, and government instabilities,
- ***economic crises*** such as extreme currency inflations or depressions, bank failures, stock market collapses, labor strikes, corporate downsizing and bankruptcies,
- ***business difficulties*** such as strikes, layoffs, corporate downsizing, bankruptcies, transportation interruptions, shortages at local stores, unavailability of raw materials or essential components,
- ***utility interruptions*** such as prolonged power outages or fuel shortages, or
- ***personal emergencies*** such as loss of employment, severe illnesses or accidents, loss of the family breadwinner, loss of the home, transfers to new locations, being sued, being imprisoned, loss of essential vehicles or tools, etc.

It is obvious that such events happen suddenly, and often without warning. A family's state of well-being, its priorities and directions, its financial status, and its prospects for survival can change radically in a brief instant.

Emergency preparedness is a philosophy, a way of life, a state of readiness that provides a buffer and and a safety shield in case such disastrous events suddenly occur.

This book is a "starter manual"—a brief and inexpensive guide for families and individuals who feel a personal necessity to prepare for possible future emergencies and difficulties. It's rooted in the famous adage, "better safe than sorry." It shows the way, gives sound advice and valuable information, and provides a strong basic ground work for maintaining stability and well-being if and when such emergencies arise.

Read it, do it, and profit from the counsel it provides.

FOOD STORAGE

Where Do I Start?

Your first task is to acquire an overview of what you need to accomplish. Read this book, and others, to start your thinking. Then begin to prepare a family emergency readiness plan—not just for food storage, but for the broader scope of emergency preparedness challenges. Write it down. Revise it as you grow, and as you progress towards your goals.

When you're ready to begin acquiring, storing and rotating your food supply, start with the basic items—wheat, milk, sugar or honey, and salt. Then work on getting the basic helpers—beans, rice, split peas, eggs, potatoes, etc.

Most of us do not have enough money to buy a whole year's supply of food at one time. Therefore, work on a three-, six-, nine-, and finally a twelve-month supply. If you bought a year's food supply of one item and nothing else, and you were required to use only that supply, your diet would be meager indeed. With a three-month supply of different items, you would at least have a variety of basic goods rather than just one item.

Plan your food-storage program and then follow your plan. This will save money and unnecessary waste of food. You are preparing yourself for a time of need, so it's best to purchase basics before obtaining luxury food items. When buying foods you haven't tried before, buy a small portion and try it before buying in bulk. Many people spend a lot of money on one item, only to find that they do not like what they've bought.

Part of your supply should be foods that you are now eating. These foods can be used to help you adjust to foods that your body is not accustomed to.

This book will help you plan your storage of food and other items you need if you refer to the charts, plans and lists included. They will help you to calculate the specific food-storage needs for your family; they will tell you what the storage requirements are for each item, and what the projected shelf-life span for each item is.

What I am going to present in this chapter is (1) a recommended storage program which I have prepared, which is designated as *Plan One*, followed by (2) several other storage lists and recommendations from other sources, and (3) several quotes which present various prospectives and philosophies on food storage. You will note that each list has somewhat different recommended items and quantities. This will allow you to approach the preparation of your own plan from several different perspectives. You'll undoubtedly want to pick and choose to match your family's personal preferences.

First, a word or two about canned goods: a 3- to 6-month supply of canned goods is recommended. These goods must be rotated to prevent spoilage. This supply may help you to adjust to the diet of wheat and other basic food storage items that might be required in a longer term situation.

How can you judge your canned goods needs? "Canned goods for a family of five will average per day, 2 soup, 2 vegetable, 1 meat, and 2 fruit. Shortening for a year will consist of 12 3-lb. cans. Juices will average 5 cans per week. Also needed are 6 cans of baking powder, 4 gallons of Clorox for water sterilization." (From *How to Prepare for the Coming Crash*, by Robert L. Preston, p. 95.)

PLAN ONE–COMPREHENSIVE

Item	Storage Life	Amount Needed Per Person Per Year

Basic Items

Wheat	indefinite	250-300 lbs

☞ *Moisture content must be less than 10%. Fumigate. Store in metal or plastic containers.*

Milk

• Non-fat powdered	5-15 yrs	80 lbs
• Instant	2-4 yrs	

☞ *To determine how much milk you need, determine how many quarts your family uses per week and multiply by 10. This will tell you how many pounds of powdered milk are needed.*

☞ *Store milk in complete darkness and below 70°.*

Salt	indefinitely	5-7 lbs

☞ *Must be kept dry.*

Sugar	indefinitely	100 lbs
Honey	indefinitely	80 lbs

☞ *Store in a cool, dry place.*

Item	Storage Life	Amount Needed Per Person Per Year
Water	indefinitely	520 gallons

☞ *Twenty-eight gallons of water per person (2 gallons per day) is a two-week emergency supply.*

☞ *Add 4 drops of Clorox per gallon.*

Basic Helpers

Beans	indefinitely	50-100 lbs

- Pinto
- Red
- Great Northern
- Pink

☞ *Keep in a cool, dry place in a sealed or air-tight container.*

Rice

• Long grain	5-10 yrs	50 lbs
• Brown	1-2 yrs	25 lbs

☞ *Keep in a cool, dry place in metal, plastic or glass containers. Fumigate.*

Macaroni	2-3 yrs	20-25 lbs

☞ *Keep in a cool, dry place free from bugs.*

Yeast	1 yr	1 lb

☞ *Keep cool and dry.*

Eggs, powdered	3-5 yrs	10 lbs

☞ *Should be kept below 40° or rotated within a year.*

Baking powder	2 yrs	3 lbs

☞ *Keep cool and dry.*

Item	Storage Life	Amount Needed Per Person Per Year
Baking soda	2 yrs	5 lbs

☞ *Keep cool and dry.*

Flour	2 yrs	50-100 lbs

☞ *Use metal containers and store in a cool, dry place. Fumigate.*

☞ *Enriched flower is best for storage.*

Potato flakes	2-5 yrs	20-30 lbs

☞ *Store in metal containers. Kept them away from moisture.*

Jell-O	2-5 yrs	10-15 lbs

☞ *Keep cool and dry.*

Green split peas	indefinitely	20 lbs

☞ *Keep in a cool, dry place. Fumigate.*

☞ *Moisture content must be less than 10%.*

Freeze-dried products	7 yrs or more	Per Family's needs and likes

☞ *Store in a cool, dry place. After opening, these foods must be used within a month. Unopened, they are good for long storage.*

☞ *Freeze-dried products may be substituted for canned goods for longer shelf life.*

Dried foods	4 yrs or more	Per Family's needs and likes

☞ *Keep containers in a cool, dry place.*

Item	Storage Life	Amount Needed Per Person Per Year
Soybean granules	5 yrs or more	50 lbs

☞ *Store in a cool, dry place in metal containers. Unflavored is best because it contains less fat. Check brands and fat content before ordering.*

Canned Goods

☞ *Canned goods should be stored in a cool, dry place, away from the furnace and water pipes.*

☞ *It is recommended that you store a 3- to 6-month supply of canned goods. These goods must be rotated to prevent spoilage.*

☞ *Canned goods can help your system adjust to a diet of wheat, powdered milk and other basic storage foods.*

☞ *Average canned goods usage for a family of five for a day will typically be two cans of soup, two cans of vegetables, two cans of fruit, and one can of meat.*

☞ *Avoid bargain sales of canned goods. They may be old stock or inferior merchandise.*

☞ *You may wish to substitute home-bottled fruits and vegetables for added savings and fresher produce.*

☞ *Items with pits will only last about one year.*

Vegetables		
• Peas	3 yrs	
• Corn	2 yrs	Store
• Beans, string	2 yrs	144 cans
• Tomatoes	2 yrs	(6 cases)
• Beets	2 yrs	per person of
• Carrots	2 yrs	assorted
• Potatoes (canned)	3 yrs	vegetables
• Spinach	2 yrs	
• Squash	2 yrs	

Item	Storage Life	Amount Needed Per Person Per Year
Fruits		
• Applesauce	2 yrs	
• Apricots	2 yrs	Store
• Berries	2 yrs	144 cans
• Cherries	1 yr	(6 cans)
• Grapefruit	2 yrs	per person of
• Peaches	2 yrs	assorted
• Pears	2 yrs	fruits
• Pineapple	2 yrs	
• Plums	1 yr	
• Prunes	1 yr	

Fats and Oils

Lard	1/2 yr	3 3-lb cans

☞ *A family of five will average about 12 3-lb cans of shortening per year.*

Margarine	1 yr	20 lbs
Vegetable oil	3 yrs	10 lbs
Corn oil	3 yrs	10 lbs
Butter	1/2 yr	

Fruit Juices

☞ *A family of five will average about five cans of juice per week.*

Tomato	1 yr	Store
Pineapple	1 yr	60 cans,
All others	1/2 yr	assorted

Item	Storage Life	Amount Needed Per Person Per Year
Meats		
Tuna (in oil)	5 yrs	
Canned beef	5 yrs	
Fish (canned)	4 yrs	3 cases, assorted
Bacon	4 yrs	
Deviled ham	4 yrs	
Spam	5 yrs	

Other Items

☞ *The storage life shown is for unopened cans and bottles.*

Item	Storage Life	Amount Needed Per Person Per Year
Peanut butter	1/2 yr	10-20 lbs
Soups	3 yrs	3 cases, assorted
Dried cheese	1 yr	
Pickles	2 yrs	
Catsup	2 yrs	as needed by the family
Spices	2 yrs	
Jams	3 yrs	
Jellies	3 yrs	
Spaghetti	4 yrs	
Bouillon	5 yrs	

Cleaning Supplies

☞ *These amounts are conservative. In times of famine, anarchy, war, or natural disasters, disease may be on a rampage. It will be essential to use extra cleaning supplies to prevent sickness.*

	Amounts Needed Per Family Per Year
Hand soap	50 bars
Cleanser	20 cans
Dish soap	3 gal
Disinfectant	2 gal

Shampoo	1 gal
Lye	10 cans

☞ *Lye can be used to make soap.*

Ammonia	2 gal
Brushes	4
Clorox	8 gal

☞ *Clorox can be used to purify water as well as for a cleaner.*

Your Food Storage Program

An excellent book in Horizon Publishers' emergency preparedness library, *Prepare Today, Survive Tomorrow*, on pp. 42-43 and 48-49, presents other valuable insights on the food storage process:

"You will need two systems of food storage to cover all contingencies: (1) an initial storage system; and (2) a continual storage system. Both are absolutely essential!

"The initial supply is food and water you have stocked through a good food-storage plan. This supply includes dehydrated, canned, and otherwise preserved foods, plus grains and sprouting seeds. It also consists of vitamins and mineral supplements you have set aside. Frozen foods require special consideration.

"Continual supply will refer to the food items that will, through proper care, replenish their own supply and be a perpetual source.

The Initial Food Storage Program

"First, consider the initial supply. A key aid will be the method by which your food is preserved.

"The most nutritious and longest-lasting method of preservation is through commercial dehydration or freeze-dried processes. Processing plants are normally located at or near farms. As delays between picking and processing time are extremely destructive to nutrient retention, farm-located factories shorten the time, which adds to nutritional values. If you can fit it in your budget, I strongly recommend that you obtain a year's supply at such a factory.

"There are many fine companies with good programs for dried food storage. The units or groups offered for designated time periods are usually well balanced, and the dried-food will provide excellent nutritional values as well as taste, flavor and variety. Don't confuse these processes with the antiquated methods of the old army chow of the 40s and 50s. Today's dried foods are tasty and pleasing.

"Freeze-dried foods are an excellent form of dehydrated foods. They have an advantage of good looks and texture over regular dehydrated foods, and there are foods that are more successfully preserved by this method. I do not encourage anyone but the very wealthy to store freeze-dried foods, primarily because they are considerably more expensive than regular dehydrated foods. They are the Cadillac of foods, but your money can be better spent on many other needs.

"I do believe that dehydrated and canned goods are an important part of your initial food stock and should be stored in sufficient quantities to be used for barter to help get your family through to first harvest.

"Many good recipe books are available for dehydrated foods, but one key point I'd like to make concerns the reconstituting of vegetables and many other items. Give them plenty of time to soak, preferably overnight, in cool, clean water. Let them take time to re-hydrate themselves naturally, without force-cooking the water back into them.

"Another thing: you are going to have to practice and use some of these foods now, to get the hang of cooking up some really delicious dishes. Don't wait until you need the food to learn how to prepare it. Buy it, use it, rotate it, and replace it. . . .

Prepared Foods Have A Place

"Another suggestion on what to incorporate in your initial supply is canned goods. Stock and rotate at least a thirty-days' supply of canned foods that do not require cooking or heating to be edible. This good suggestion is also made in our civil defense programs. Since our plight could very well be a result of a nuclear attack, we need to be ready for a transitionary period to put our regular supply into effect. Power may be out, gas supplies may be disrupted, and we will need a food available that can be prepared without much water or heat."

PLAN TWO—THE BASIC FIVE

This plan, and the material which introduces it, is adapted from *Prepare Today, Survive Tomorrow*, by Joe Harold, pp. 49-52, and is used by permission from the publisher, Horizon Publishers (Bountiful, Utah). This book, by the way, presents a wide perspective of very valuable emergency readiness information, even dealing realistically with principles of personal self defense in times of social collapse.

"As you start accumulating you may find the expense a little difficult to manage.

"It is for this reason that I present the basic five as a good alternative that any serious-minded person can afford. It is actually what I prefer as an available second-year supply in case you aren't able to replenish your regular supply at the end of the first year. . . .

"The basic five include:

1. wheat
2. powdered milk
3. salt
4. honey
5. vitamin-mineral supplements

"To use this as a first-year accumulation would be acceptable, but not quite as pleasant. If this most-economical approach is the avenue you take, start on the program as soon as possible and incorporate as many dehydrated foods as you can. At least expand the basic five to include sprouts . . .

"Ground wheat can be mixed with water and washed clean, leaving a gluten base which can be used as a meat substitute. I have broiled gluten in beef broth, flavored it, and fried it like chicken-fried steak. We've also run it through a meat grinder to make meat balls and meat loaf. I have also used it as a beef base in soups and stews. It's really great! Of course, you can also use this gluten when you bake your favorite bread and cookies. Ground wheat can be prepared for a delicious cereal, or whole steamed wheat will make a nice rice or potato substitute.

"Sprouting wheat will literally explode its value in vitamin C and B. It provides the necessary amino acids for good health as well as vitamin E. Wheat sprouts are sweet and delicious eaten as sprouts, or they can be cooked into breads, soups, meatloafs, and cereals, as already mentioned.

"In figuring your needs, store about 600 pounds per person, half to be eaten and half to trade or plant. One five-gallon plastic bucket will hold about 35 pounds of wheat. When you store your wheat, put about three tablespoons of crushed dry ice in the bottom of the bucket, then fill to about one inch from the top with recleaned wheat available at all big grain mills. Be sure to specify a hard winter wheat, 10 percent or less in moisture content, with 12- to 14-percent protein. The mill will have this information for you. Don't buy hybrids because they may not reproduce if planted.

"Before you snap the lid on your bucket, just lay the lid loosely in place for thirty to forty-five minutes. Let the dry ice evaporate and fill the bucket. You can't see CO2, but it's heavier than air so the excess will just spill out. Now gently, without stirring or removing the lid, snap it in place. Watch the bucket closely for a while. If it starts to swell the lid, just raise the edge and "burp" the bucket. Then seal it back until it stops swelling. If you don't do it this way the lid will blow off under pressure and will scatter wheat for twenty feet. The dry ice, or CO2, will kill any little varmints that are still in egg form. I've seen those little creatures literally eat a third of a bucket of wheat in less than a year.

"Treat your wheat this way and it'll keep and still sprout, even for your great-great-grandchildren. Wheat has been documented to have still sprouted after 2,500 years storage in one of the old pyramids.

Wheat is great stuff to trade, eat, or switch to your continual supply.

"Now, let's turn to item number two of the 'basic five.' Purchase and package one hundred pounds of powdered milk for each person in your family. Instant milk mixes easier, but non-instant stores better and tastes better. Put away a hand-powered egg beater to help you mix it. I recommend non-instant milk, which can usually be bought in quantity form from a dairy in your area. You can buy fifty-pound bags very economically, and some in No. 10 cans sealed with nitrogen.

"After you find your source for fifty-pound bags, remember this: milk must be stored in a cool, dry, dark place—all of these requirements must be met, not just one or two!

"What I do is line a plastic or tin bucket or tin with a black plastic trash bag of whatever size I want. Pour the milk powder in carefully sealed smaller bags. Put these bags in the bucket, tie the black bag, seal the container, and store it in a cool, dry place. The milk will keep for three to four years, easily. You can open the large bag, remove one little bag at a time, and leave the others for later. Rotate the milk and use it regularly. After you've reconstituted the milk, let it sit overnight in a refrigerator or cooler to enhance the flavor.

"The third step in this program can be completed by filling two or three plastic buckets with the little one-pound spout-type boxes of iodized and non-iodized salt. That way you'll have plenty for yourself as well as another good trade item. You might also go to a feed store and buy a couple of the salt blocks cows lick on. These little items are super-cheap now and can return investments of over 1000 percent when traded.

"The fourth phase will require storing a five-gallon bucket of pure honey for each member of the family. If you can handle the diet, you can all live for one year without using a single can of your luscious dehydrated food. The same 2,500-year-old pyramid story includes honey that was still edible.

"Finally, you might find your health supported better, even over your mistakes in using your supply, if you will include a 500-tablet bottle of one-per-day multiple vitamins and minerals for each person. Just about any brand of vitamins will do. You won't be totally dependent on them anyway, especially if you sprout a lot of wheat.

"After you have provided yourself with the basic five items, start making them more palatable. You must get a wheat grinder, sprouting trays, a good sprout book, an assortment of sprouting seeds, and a variety of dehydrated foods. The dehydrated foods you add will make your meals more enjoyable, and for some of you they might even be essential. Other items you will enjoy are sugar, coffee, yeast, hydrogenated oil, pepper, garlic powder, and chili powder. I could go on and on, but you can use your imagination. Just don't sell sprouts short. If you don't "get into" sprouts, you won't have anything fresh, alive, and green for a year or more. They are extremely important!

"The following chart will specify more accurately the precise amounts of storage foods required for each family member by sex and age. My general recommendations exceed actual need in order to provide excesses for trading, loss, surprise guests, and possible neighborhood needs. If your desire is to store the absolute minimums then go by this chart."

Food Storage Item	Quantity in Pounds Required for Each Family Member (Rounded to nearest 5 pounds)									
	Adults		Children				Teenagers			
							13-15 yrs		16-20 yrs	
	Male	Female	1-3	4-6	7-9	10-12	Girl	Boy	Girl	Boy
Wheat	300	200	70	100	160	210	220	280	200	335
Nonfat dried milk	55	45	100	100	100	100	100	100	100	100
Sugar and/or honey	30	20	15	20	25	30	30	40	30	50
Salt	5	5	5	5	5	5	5	5	5	5

As you can see, *Prepare Today, Survive Tomorrow* provides excellent information on using the basic five foods.

PLAN THREE
An Expanded Basic Food Storage Program

Another very popular food storage program, which has been widely circulated in the Intermountain area, expands on the basic five ingredients while still utilizing foods with a long storage life. It offers better variety, and provides realistic quantities based on gender and age groupings:

Recommended Food Storage for 12 Months for Various Age Groups, by Gender

	Unit	Male (over 18)	Male (7-18)	Female (over 18)	Female (12-18)	Female (7-11)	Child (under 7)
Grains Group							
Wheat	lbs	189	176	132	139	151	107
Flour, enriched white	lbs	17	16	12	12	14	10
Corn meal	lbs	42	39	29	31	34	24
Oats, rolled	lbs	42	39	29	31	34	24
Rice, white enriched	lbs	84	78	59	62	67	48
Barley, pearled	lbs	4	4	3	3	3	2
Spaghetti & ***macaroni***	lbs	42	39	29	31	34	24
Totals		**420**	**391**	**293**	**309**	**337**	**239**
Legumes Group							
Beans (dry)	lbs	25	25	25	25	25	25
Beans, lima (dry)	lbs	1	1	1	1	1	1
Beans, soy (dry)	lbs	1	1	1	1	1	1
Peas, split (dry)	lbs	1	1	1	1	1	1
Lentils (dry)	lbs	1	1	1	1	1	1
Soup mix (dry)	lbs	5	5	5	5	5	5
Totals		**34**	**34**	**34**	**34**	**34**	**34**
Fats & Oils Group							
Vegetable Oil	gal	2	2	2	2	2	2
Shortening	lbs	4	4	4	4	4	4
Mayonnaise	qts	2	2	2	2	2	2
Salad Dressing (mayonnaise type)	qts	1	1	1	1	1	1
Peanut butter	lbs	4	4	4	4	4	4
Totals		**13**	**13**	**13**	**13**	**13**	**13**
Milk Group							
Milk, nonfat (dry)	lbs	14	14	14	14	14	14
Evaporated milk (12 oz. net wt.)	cans	12	12	12	12	12	12
Totals		**26**	**26**	**26**	**26**	**26**	**26**
Sugars Group							
Sugar, granulated	lbs	40	40	40	40	40	40
Sugar, brown	lbs	3	3	3	3	3	3
Molasses	lbs	1	1	1	1	1	1
Honey	lbs	3	3	3	3	3	3
Corn syrup	lbs	3	3	3	3	3	3

Jams & preserves	lbs	3	3	3	3	3	3
Fruit drinks, powdered	lbs	6	6	6	6	6	6
Gelatin, flavored	lbs	1	1	1	1	1	1
Totals		**60**	**60**	**60**	**60**	**60**	**60**
Miscellaneous Group							
Yeast, dry	lbs	1	1	1	1	1	1
Soda	lbs	1	1	1	1	1	1
Baking powder	lbs	1	1	1	1	1	1
Vinegar	gal	1	1	1	1	1	1
Chlorine bleach	gal	1	1	1	1	1	1
Salt, iodized (8 lbs/person/year)	lbs	8	8	8	8	8	8
Water (28 gal/person/ for 2 weeks)	gal	28	28	28	28	28	28
Totals		**41**	**41**	**41**	**41**	**41**	**41**

Using Your Computer to Inventory Your Storage Supplies

Tracking your food storage in your computer is the way to go these days! An inventory system is easy to set up in a basic spreadsheet program like *Excel*, *Lotus 1-2-3*, or *Quattro Pro*. Such a system has many advantages. It is easy to expand or contract: you can add or delete products anywhere in the spreadsheet. Likewise, it's easy to print out a sheet and use it for taking a monthly or quarterly inventory of your supplies on hand. You can use it to track how much investment has been made in the stored items, and use the same format to estimate and budget what future storage items will cost. If you want a more sophisticated record, you can record the date purchases are made, and use those dates to help in your food rotation procedures. If additional quantities of a listed item are purchased at a different price, an extra line item can be inserted with the different quantities and values. You may opt to italicize the older item, indicating that the food it still lists as available should be eaten before you begin to use the newer items. When it is all eaten, delete the italicized line. You can make a column showing what the total quantity purchase goal is for each item, and you can have the computer calculate how many more units need to be purchased to reach the goal. If you wish, you can make the computer calculate the cost of the planned future purchase. You can make additional files in which you assign a priority order to future purchases, planning what month you wish to make them. In short, you can make your computer do just about everything except cook and eat the stored foods.

The chart below was devised by Duane S. Crowther (who edited and published this book), as part of his assignment as an Emergency Preparedness Chairman in his local area. The example shows appropriate quantities for a 12-month supply for a family of four members, based on 1998 prices in the Salt Lake City, Utah area. You are invited to adapt it for your personal needs, and to apply any or all of the suggestions which he made in the preceding paragraph.

	Quantity	*Unit Cost*	*Extended Cost*	*Unit*	*Typical Packaging*
GRAINS GROUP					
Wheat	480	0.159	76.32	lb	50 lb: $7.99
Whole wheat flour	60	0.147	8.82	lb	25 lb: $3.69; 5 lb, 10 lb
Enriched white flour	48	0.147	7.06	lb	25 lb: $3.69; 5 lb, 10 lb
Corn (canned: whole kernel)	60	0.450	27.00	can	16 oz: $.45
Corn meal	60	0.394	23.64	lb	5 lb: $1.97
Oats, rolled	120	0.539	64.68	lb	50 lb: $26.98
Rice, white enriched	240	0.315	75.60	lb	25 lb: $7.89
Pearled barley	12	0.890	10.68	lb	16 oz: $.89; 25 lb: $8.99
Spaghetti & macaroni	120	0.863	103.56	lb	3 lb: $2.59
Total	**1,200**		**$397.36**		

	Quantity	*Unit Cost*	*Extended Cost*	*Unit*	*Typical Packaging*
LEGUMES GROUP					
Beans (dry) {small Red}	252	0.599	150.95	lb	25 lb: $14.98
Beans, lima (dry)	12	0.799	9.59	lb	25 lb: $19.99
Beans, soy (dry)	6	0.519	3.11	lb	25 lb: $12.98
Peas, split (dry)"	15	0.300	4.50	lb	25 lb: $7.49; 2 lb: $.69
Lentils (dry)	15	0.419	6.29	lb	25 lb: $10.49
Total	**300**		**$174.44**		
FATS & OILS GROUP					
Vegetable oil	19	5.490	104.31	gal	1 gal: $5.49
Shortening	49	0.830	40.67	lb	3 lb: $2.49
Mayonnaise	8	1.890	15.12	qt	1 qt: $1.89
Salad dressing (mayonnaise type)	12	1.990	23.88	qt	1 qt: $1.99
Peanut butter	39	1.490	58.11	lb	3 lb: $4.49
Total	**127**		**$242.09**		
MILK GROUP					
Milk, nonfat dry	80	1.170	93.60	lb	50 lb: $58.98
Evaporated milk (12 oz net wt)	60	0.430	25.80	can	1 12 oz can: $.43
Cheese (dehydrated)	10	4.600	46.00	lb	3 lb,4 oz: $14.98
Total	**150**		**$165.40**		
SUGARS GROUP					
Sugar, granulated	402	0.359	144.32	lb	25 lb: $8.99
Sugar, brown	30	0.535	16.05	lb	2 lb: $1.07
Molasses	6	2.320	13.92	lb	24 oz: $3.49
Honey	30	0.998	29.94	lb	5 lb; $4.99
Corn syrup	30	1.990	59.70	qt	1 qt: $1.99
Jams & preserves	30	1.190	35.70	lb	32 oz: $2.39
Fruit drink, powdered	60	1.590	95.40	lb	39 oz: $3.87
Flavored gelatin	12	2.100	25.20	lb	6 oz: $.79
Total	**600**		**$420.23**		
MISCELLANEOUS GROUP					
Dry yeast	2	1.995	3.99	lb	2 lb: $3.99
Soda	4	0.530	2.12	lb	1 lb: $.53
Baking powder	4	2.160	8.64	lb	14 oz: $1.89
Vinegar	2	1.790	3.58	gal	1 gal: $1.79

	Quantity	Unit Cost	Extended Cost	Unit	Typical Packaging
Chlorine bleach	1	1.130	1.13	gal	1 gal: $1.13
Salt (iodized)	20	0.170	3.40	lb	2 lb: 3/$1.00
Water	112	0.000	0.00	gal	(28 gal/person/2 wks)
Total	**145**		**$22.86**		
					Disk: C,L,1
					File: Foodcost.Wk4
Grand Totals	**2,522**		**$1,422.37**		Range: A1..L70

Food Storage: The Planned, Sensible Accumulation of Goods

Certainly, one of the most useful books on emergency preparedness available today is *Project Readiness: A Guide to Family Emergency Preparedness*, by Louise E. Nelson. The book is thorough, filled with useful data, often quoted, and has long held a respected place in the food-storage and preparedness industries as being an essential, "must have" source. The author presents some valuable insights on food storage as well as several storage lists and plans. Again, with the permission of Horizon Publishers, I will quote several of these insights and plans. All quoted items are from *Project: Readiness*, pp. 43-52.

One very valuable contribution is the following definition of food storage:

Storing is the planned, sensible accumulation of goods during times of plenty to prepare for times of need.

This is the author's (Louise Nelson's) commentary:

"The idea of a home storehouse is not new. For most of us it is a relic of bygone days. When our economy was largely agricultural, it was the established practice for families to set aside food from one year's harvest to feed the family until the next harvest. Before the age of the supermarket, a housewife filled her fruit cellar with home-canned fruits and vegetables. The hog was slaughtered, smoked and hung in the smokehouse all winter, the meat being used as needed.

"There is a difference between storing and hoarding. Hoarding is the greedy gathering of material belongings, including food, without plan or good reason. Hoarding is the result of panic. *Storing is the planned, sensible accumulation of goods during times of plenty to prepare for times of need.*

"The farm family with a freezer full of home-grown beef and full root and fruit cellars is not accused of hoarding, but praised for wisdom and forethought. The urban family with a full pantry of either home-canned or commercially-processed foods and a full freezer should be considered equally wise.

"Panic buying does not prevent price increases. On the contrary, by producing false shortages, it drives prices higher.

"No one can devise the perfect storage plan that will suit the needs of every individual or family. Therefore, the plans in this book are meant to serve only as guides. You will be able to determine from them the types of goods and quantities that some families have found to be adequate for their needs.

"The first step in determining what to store is to decide whether you want a storage plan that will require *continual rotation*, or one that will require *infrequent rotation*, or a *combination of the two.* Most families will find that a combination plan offers the greatest amount of flexibility.

Long-term Storage

"This type of plan relies heavily on wheat, beans and other dried seeds and grains. Also included in a long-term storage plan are the low-moisture, dehydrated foods.

"The relatively small amount of storage space required for these items in comparison with wet-pack foods is the principle advantage of this plan. Also, these foods require little maintenance and can be held in storage for extended periods with little deterioration.

"The low cost of these foods is a definite advantage. One recent comparison of cost-per-serving of low-moisture and wet-pack canned foods shows that the low-moisture foods cost substantially less. Because of regional variations in prices and the present inflationary trend in food prices, you should make your own comparison, taking into account any waste in fresh foods or wet-packed canned foods.

"The major disadvantage is that few of us are accustomed to this type of diet. Also, large amounts of water are needed to prepare these foods. In the event of a major disruption of the water supply, these foods would be virtually useless without having significant quantities of water stored. Once a can of low-moisture food is opened, the contents must be used within a certain period of time to prevent deterioration. Beans, wheat and other such products are quite inexpensive and pack a great deal of nutrition. However, one must accustom his body to eating large quantities of these foods.

Rotation Storage

"This type of plan calls for the storage of canned foods, fresh fruits and vegetables, and frozen foods.

"The major advantage of this storage program is that one can store the foods he is accustomed to eating. Stores of large quantities of water are not needed in order to make good use of these goods. Most canned foods are packed with water in some form. The water in them can be used for drinking and preparation of foods that require the addition of water.

"It is the water in these foods that constitutes the major disadvantage, however. Because of weight and bulk, they require more space and are harder to move than the long-term storage foods. There is a certain amount of maintenance needed to keep these foods in good condition. Most must be rotated periodically, and some foods must be inverted or shaken at regular intervals to maintain quality. Attention must be paid to keeping your stock fresh. With this type of storage plan, you cannot just purchase the supply, store it and forget it.

Combination Plan

"Let us assume that your goal is to have a one-year's supply of food on hand. First, purchase a six-months' supply of long-term storage items. Then purchase a good variety of foods that require rotation.

"This way you will have a good supply of nourishing and inexpensive foods available while you are purchasing the more costly foods that require rotation. By incorporating the rotated foods in your daily meal planning you can keep them in constant rotation. Regularly including the long-term storage items in your meals will help your families to become accustomed to eating those foods.

"Now, after you have decided which type of a storage program to use, the next step is to decide what and how much to store.

"To determine how much will be needed for a rotation program, the best method is to keep a record of everything you use for a certain period of time, such as two weeks. Then multiply that to obtain the amount needed for a year or whatever period of time for which you desire to maintain stores. This requires very detailed records of everything used, right down to the 1/2 teaspoon of soda used in Sunday's pancakes. Place a chart on the refrigerator or some convenient place and record what you use as you use it. Be sure to include all exhaustible supplies such as food, soap, paper products, medicines, light bulbs and baby supplies. Then inventory your closets and cupboards to determine what additional supplies you may need that were not used during the period for which you kept your record. While taking your inventory, record what supplies you have on hand.

Some Tips on Storage

"The prime rule of food storage is: ***store what you use and use what you store.***

"Test all products before purchasing large quantities for storage. If you don't like a product, or you can't eat it or wear it, it will do you no good to have it on the shelf.

"Tests have shown that a person will sometimes starve himself rather than eat an unfamiliar food. This is especially true of children. Serve the foods that you store so that your family will be accustomed to them. There will be fewer problems maintaining good health on your storage diet. When you throw away 20 cents worth of food, you might as well be throwing away 20 cents in cash. It amounts to the same thing.

"Purchase basic items first. Then add frills, if desired. Always keep in mind the goal is to be as adequately prepared as possible at all times, should an emergency arise.

"Make a month-by-month purchasing plan that will maintain adequate balance among the foods that you have stored so that you can use your storage conveniently and have proper nutrition and variety at any time while you are building up your supply.

"Plan for variety. Monotony can lead to loss of appetite and eventually to malnutrition from improper eating.

"Don't forget to include in your storage plan the supplies that you have on hand for current use.

"Take advantage of sales. But be aware that many items placed on sale may be old stock that is being cleared to make way for new. Sales of canned fruits and vegetables during the summer may be just such a clearance. Allowances for the time already stored before purchase must be made in determining the shelf life of the product.

"Shelf life as given in tables is not absolute. Most foods will keep much longer. The tables will help you to use your food while it has the best quality and the highest nutrition. Do not feel that food has to be thrown out if it has been stored longer than the table advises.

"Inspect your storage frequently to avoid costly losses from insects, rodents, rusting cans, water seepage into the storage area, or other causes.

"Avoid dented or rusted cans, regardless of cost. These should never be purchased for storage. Likewise, be wary of merchandise on the merchant's half-price table. These items may be near the end of their storage life. If in doubt, ask the manager why the item has been reduced in price.

"Buy the item that will give you the most nutrition for the least price. Regular low-moisture foods will give you many more servings per dollar than will freeze-dried foods. However, some foods are available only in the freeze-dried form. Where you have a choice, regular low-moisture food is generally the better buy. Always compare costs per serving.

"Similarly, dried beans cost much less per serving than do cooked, canned beans. The basic ingredients for baked goods or casseroles are less costly than the prepared mixes. When you consider the total cost of a year's supply of food, those pennies saved on each item can add up to a lot of dollars on the total supply. . . .

"Any storage plan should be designed to give adequate nutrition. The USDA Food Pyramid will serve as a guide."

The USDA Food Pyramid

Suggested servings of each food group needed daily by an individual for good nutrition

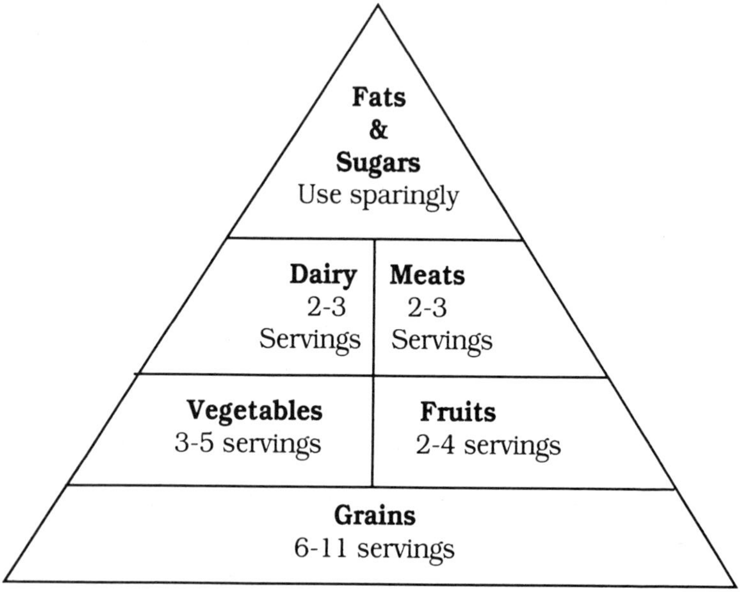

What constitutes a serving of each food group?

Dairy:
One cup of milk or yogurt, $1\frac{1}{2}$ oz, natural cheese

Meats:
Three oz. lean meat, 1 egg, 2 tablespoons peanut butter, $\frac{1}{2}$ cup dry beans

Vegetables:
One cup raw leafy vegetables, $\frac{3}{4}$ cup juice, $\frac{1}{2}$ cup cooked vegetables

Fruits:
One med. apple, orange or banana, $\frac{1}{2}$ cup chopped fruits, $\frac{3}{4}$ cup juice

Grains:
One slice bread, 1 oz. cereal, $\frac{1}{2}$ cup rice or pasta

PLAN FOUR

A Personal Storage Plan For One Year

The following very comprehensive plan is presented by Louise Nelson in her book, *Project: Readiness*, pp. 50-54, and is cited here by permission of Horizon Publishers:

Plan I feeds 1 adult, or 1 teenager, or 2 small children
Plan IV feeds 4 adults, or 4 teenagers, or 8 small children

DH = *Dehydrated*
D = *Dried*

	Amount Needed	
	Plan I	**Plan IV**
Meats And Other Protein Foods		
Salmon	6 (1-lb.) cans	26 (1-lb.) cans
Ham, canned or frozen	12 lbs.	45 lbs.
Bacon, canned	10 lbs.	39 lbs.
Chicken, canned or frozen	6 whole	26 whole
Luncheon meat	13 cans	52 cans
Corned beef	6 cans	26 cans
Tuna	20 cans	78 cans
Roast beef, canned	6 cans	26 cans
Stew beef, frozen	7 lbs.	26 lbs.
Roasts, frozen	10 lbs.	40 lbs.
Ground beef, frozen	10 lbs.	40 lbs.
Dried beef	13 lg. jars	52 lg. jars
Sausage, canned or frozen	6 lbs.	26 lbs.
Textured Vegetable Protein (TVP), *DH*	1 (#10) can	1 (#10) can
Pork and beans	13 (1-lb.) cans	52 (1-lb.) cans
Chili with beans	6 (1-lb.) cans	26 (1-lb.) cans
Limas, *DH*	6 lbs.	26 lbs.
Lentils, *DH*	6 lbs.	26 lbs.
Soybeans, *DH*	12 lbs.	45 lbs.
Pinto beans, *DH*	14 lbs.	55 lbs.
Eggs, *DH*	3 (#10) cans	11 (#10) cans
Milk and Milk Products		
Evaporated milk	18 lg. cans	70 lg. cans
Pineapple cheese spread	6 jars	26 jars

	Amount Needed	
	Plan I	**Plan IV**
Cheese	16 lbs.	65 lbs.
Powdered milk	70 lbs.	280 lbs.
Pasta		
Macaroni	4 lbs.	15 lbs.
Noodles	6 lbs.	25 lbs.
Rice	10 lbs.	40 lbs.
Fats		
Shortening (for baking and frying)	9 (3-lb.) cans	35 (3-lb.) cans
Margarine (for eating)	6 lbs.	26 lbs.
Peanut butter	1-3 lg. jars	6-12 lg. jars
Fruits		
Pie fillings	6 cans	26 cans
Grapefruit segments, canned	6 cans	26 cans
Assorted canned fruits	40 cans or 20 quarts	156 cans or 78 quarts
Fresh apples	1/2 bushel	2 bushels
Apple granules, *DH*	1 (#10) can	1 (#10) can
Apple slices, *DH*	1 (#10) can	3 (#10) cans
Prunes, *DH*	xxxxx	2 (#10) cans
Fruit blend, *DH*	1 (#10) can	2 (#10) cans
Peach slices, *DH*	1 (#10) can	2 (#10) cans
Date bits, *DH*	xxxxx	1 (#10) can
Banana slices, *DH*	xxxxx	1 (#10) can
Juice or juice substitute	54 quarts	215 quarts
Raisins, *D* or *DH*	4 lbs.	20 lbs.
Sugars		
Granulated (for cooking and eating)	37 lbs.	150 lbs.
Granulated (for canning)	25 lbs.	100 lbs.
Powdered sugar	6 lbs.	25 lbs.
Brown sugar	6 lbs.	25 lbs.
Honey	2 lbs.	10 lbs.
Flour and Grains		
Flour or wheat for baking	100 lbs.	400 lbs.
Wheat, for cereal, etc.	10 lbs.	40 lbs.

	Amount Needed	
	Plan I	**Plan IV**
Oatmeal	4 lbs.	15 lbs.
Cornmeal	4 lbs.	15 lbs.
Rolled wheat	1 lb.	5 lbs.
Cracked wheat	3 lbs.	10 lbs.
Wheat germ	1 lg. jar	3 lg. jars
Rye flour	4 lbs.	15 lbs.
Vegetables		
Tomatoes, canned	45 (1-lb.) cans	182 (1-lb.) cans
Tomato sauce	13 cans	52 cans
Enchilada or taco sauce	4 cans	12 cans
Canned vegetables, assorted	26 cans	104 cans
Green beans, canned	6 cans	26 cans
Corn, canned	7 cans	26 cans
Spinach, canned	13 cans	52 cans
Yams, canned	6 cans	26 cans
Potatoes, fresh	15 lbs.	50 lbs.
Hash browns, *DH*	1 (#10) can	2 (#10) cans
Mashed potatoes, *DH*	1 (#10) can (173 servings)	3 (#10) cans (519 servings)
Onions, fresh	5 lbs.	25 lbs.
Carrots, fresh	16 lbs.	65 lbs.
Assorted vegetables, *DH*	2 (#10) cans	8 (#10) cans
Cabbage, *DH*	1 (#10) can	1 or more (#10) cans
Celery, *DH*	1 (#10) can	1 (#10) can
Green beans, *DH*	1 (#10) can	1 (#10) can
Soups		
Vegetable	13 cans	52 cans
Mushroom	6 cans	26 cans
Tomato	6 cans	26 cans
Cream of chicken	7 cans	26 cans
Adjuncts		
Assorted seasoning & sauce mixes	12 pkgs.	48 pkgs.
Yeast	2 cups	8 cups
Molasses	2 quarts	8 quarts

	Amount Needed	
	Plan I	**Plan IV**
Seasonings:		
Cinnamon		
Maple Flavoring		
Vanilla		
Italian seasoning		
Caraway seed		
Chili powder		
Salt (for seasoning and canning)	6 to 10 lbs.	25 to 35 lbs.
Pepper		
G. Washington Broth™	2 pkgs.	8 pkgs.
Bouillon	1 sm. jar	1 lg. jar
Mustard	1 pint	2 quarts
Garlic powder	1 sm. jar	2 lg. jars
Dry mustard	1 sm. can	1 lg. can
Pudding mix	13 pkgs.	52 pkgs.
Sour cream	6 sm. cans	26 sm. cans
Chocolate chips	6 pkgs.	26 pkgs.
Mayonnaise	4 quarts	13 quarts
Gelatin desserts	13 pkgs.	52 pkgs.
Plain gelatin	6 envelopes	26 envelopes
Vinegar	1 pint	2 quarts
Nuts for cooking and snacks	4 lbs.	15 lbs.
Pickles and relishes	3 jars	12 jars
Cocoa	1 can	3 cans
Coconut	1/2 lb.	2 lbs.
Catsup	3 lg. bottles	12 lg. bottles
Baking powder	1-2 lb. cans	3-6 lb. cans
Baking soda	1-2 lbs.	3-6 lbs.
Vitamin C tablets	365 tablets	1460 tablets
Nice To Have		
Condensed milk	3 cans	12 cans
Mushrooms	6 cans	24 cans
Gravy mixes	12 pkgs.	50 pkgs.
Popcorn	1-2 cans	6 cans
Flavored drink mixes	25 pkgs.	100 pkgs.
Candy	3 pkgs.	12 pkgs.

	Amount Needed	
	Plan I	Plan IV
Corn starch	1-2 boxes	6 boxes
Whipped topping mix	6 boxes	24 boxes
Worcestershire sauce	1 bottle	2 bottles
Soy sauce	1 bottle	2 bottles
Salad dressings	3 bottles	12 bottles
Olives	3 cans	12 cans
Chow mein noodles	3 cans	12 cans
Green pepper, DH	1 (#10) can	1 (#10) can

Plan IV would cost approximately as much as the average family of 4 or 5 spends on a year's groceries.

PLAN FIVE

A Suggested Storage Plan Based on Wheat, Honey, Salt, and Powdered Milk

The following plan is based primarily on information found in *Passport to Survival,* by Esther Dickey (Salt Lake City, Utah: Bookcraft, Inc., 1969).

Amount needed for one person per day, for a balanced diet:

- 1-1/2 oz. honey
- 1 lb. wheat
- 1 qt. reconstituted non-fat milk
- salt
- Vitamin C—from wheat sprouts or greens or vitamin tablets

This requires these amounts per adult per year:

- 365 lbs. wheat
- 35 lbs. honey
- 60 lbs. powdered milk
- 5 lbs. or more of salt
- 365 vitamin C tablets

Variation: For 2300 calories per day per adult store:

- 300 lbs. wheat
- 100 lbs. powdered milk
- 100 lbs. sugar
- 5 lbs. salt
- 365 vitamin C tablets

PLAN SIX

A Suggested Food Storage Plan Using Dehydrated Foods

The following plan is adapted from Chart 6, p. 17, in the book, *Just Add Water*, by Barbara D. Salsbury (Bountiful, Utah: Horizon Publishers). Used by permission of Horizon Publishers.

Item	Number in Family									
	1	2	3	4	5	6	7	8	9	10
Wheat—5-gal. cans	5	10	15	20	25	30	35	40	45	50
Milk—quarts	365	730	1095	1460	1825	2190	2555	2920	3185	3650
Gelatin—24-oz. cans	4	8	12	16	20	24	28	32	36	40
Eggs—1-lb. cans	3	6	9	12	15	18	21	24	27	30
Bouillon—1-lb cans	2	4	6	8	10	12	14	16	18	20
Cereal—No. 10 can	4	8	12	16	20	24	28	32	36	40
Peas—No. 10 can	5	10	15	20	25	30	35	40	45	50
Beans—No. 10 can	5	10	15	20	25	30	35	40	45	50
Rice—No. 10 can	5	10	15	20	25	30	35	40	45	50
Potatoes—No. 10 can	4	8	12	16	20	24	28	32	36	40
Vegetables—No. 10 can	4	8	12	16	20	24	28	32	36	40
Stew—No. 10 can	4	8	12	16	20	24	28	32	36	40
Onions—No. 10 can	2	4	6	8	10	12	14	16	18	20
Tomato crystals—No. 10 can	2	4	6	8	10	12	14	16	18	20
Fruit—No. 10 can	5-6	10	15	20	25	30	35	40	45	50
Juice crystals—24 oz.	4	8	12	16	20	24	28	32	36	40
Sugar or honey—lbs.	60	120	180	240	300	360	420	480	540	600
Salt—lbs.	5	10	15	20	25	30	35	40	45	50

PLAN SEVEN
One Year's Supply
Of Low-moisture Foods

The following is a plan suggested by Perma-Pak, a low-moisture foods distributor in Salt Lake City, Utah:

Suggested supply for one adult for one year:

Fruits

1 #10 Raspberry Apples
1 #10 Strawberry Apples
1 #10 Apple Slices
1 #10 Apple Sauce
1 #10 Fruit Galaxy
1 #10 Banana Chips
1 #10 Apple Drink

Meat Substitutes (A Tasty Soy Product—No real meat)

1 #10 Beef
1 #10 Bacon
1 #10 BBQ
1 #10 Chicken
1 #10 Taco
1 #10 Sausage

Vegetables

2 #10 Soup & Stew Blend
1 #10 Green Beans
1 #10 Green Peas
1 #10 Diced Carrots
1 #10 Potato Slices
2 #10 Potato Granules
1 #10 Potato Dices
1 #10 Onion
1 #10 Sweet Corn
1 #2/12 Tomato Powder

Grains, Cereal & Legumes

Grains in Metalized Bags, Sealed in Heavy-Duty Buckets

2 5-lb Spaghetti
2 5-lb White Rice
1 5-lb Pearl Barley
2 4-lb Macaroni
1 5-lb Chili Beans
1 5-lb Split Peas
1 5-lb Navy Beans
1 5-lb Yellow Cornmeal
2 5-lb Popcorn
2 4-lb Rolled Oats
3 4-lb Multi-Grain Cereal
1 5-lb Lentils
1 5-lb Small White Beans
1 5-lb Pinto Beans
5 45-lb Hard Wheat

Dairy, Cooking & Supplements

6 #10 Regular Non-fat Milk
1 #10 Margarine Powder
1 #10 Cheese Blend
2 #10 Egg Mix
1 #2/12 Soup Bouillon
1 #2/12 Corn Starch
1 Bucket White Sugar

PLAN EIGHT

Four Month's Supply Of Low-moisture Foods For A Family Of Five

The following plan is recommended by *Sam-Andy*, a low-moisture foods distributing firm based in Beaumont, California:

6 No. 10 cans fruits
12 No. 10 cans cereals and grains
3 cornmeal (13.5 lbs.)
3 oatmeal (8.25 lbs.)
3 rice (16.5 lbs.)
2 elbow spaghetti (7.5 lbs.)
1 popcorn (5.5 lbs.)
6 No. 10 cans protein foods
2 whole egg solids
1 chicken chunks
1 red beans (5 lbs.)
1 white beans (5 lbs.)
1 split peas (5.5 lbs.)
24 cans powdered milk (54 lbs.)
12 No. 10 cans vegetables
1 cabbage
4 potato granules
1 carrots
1 potato dices
1 corn
1 peas
1 onion slices
1 tomato flakes
1 celery
165 lbs. wheat or flour

Consult your distributor of low-moisture foods for more detailed and varied plans.

ROTATION

Reasons for Rotating Stored Foods

One of the fundamental aspects of successful quantity buying is rotation. It is mandatory that we rotate our foods for the following reasons:

1. To prevent spoilage and to minimize loss of food value.

2. To keep our taste buds familiar with the foods that we have purchased. This is particularly true with the basic items such as wheat, rice and beans. The only way we can be certain that the basic foods will be eaten is to include them in our diet now. If you have purchased for storage only, and don't use it now, then you'll probably never use it. Tests have shown that people will not eat what they don't like. They would rather go hungry. Purchase only those items you'll eat.

3. To prevent the loss of money and savings. A well-planned storage program and a great savings on your food bill can be easily destroyed by improper rotation. You can wipe out your entire savings program by not carefully rotating your supplies.

Canned goods that have been stored for a long period of time can safely be used, provided the cans do not bulge. The temperature of the storage room is an important factor in how well canned goods retain their nutritional value and their safety. Studies show that the rate at which chemical reaction occurs doubles with each 18 degrees Fahrenheit that temperature rises. Fifty degrees F. storage will preserve canned foods four times longer than 80 degrees F. temperature.

You should expect shorter shelf life on pitted- and high-acid fruits such as grapefruit, cherries, plums and orange juice.

To aid in rotation, the chart in the next chapter will help. The first steps show the average life span of a given product. These lists will in no way guarantee that these products will keep this long. It depends upon your storage area. The sketch on page 49 shows a type of shelf that is self rotating. If this method is not appropriate for your storage area, the following steps should be considered:

1. Date the can when you purchase it.

2. Use the oldest date first.

3. Turn the cans over every three months to prevent settling and separating.

AVERAGE SHELF LIFE OF CANNED FRUITS AND VEGETABLES

This chart shows the average shelf life of canned goods stored in the Western United States at 70° F. (21°C.). Foods stored at 50° F. will have a much longer shelf life; while those stored at 80° F. will have a much shorter shelf life.

Product	Shelf Life[1] (in months)	Product	Shelf Life (in months)
	Fruits		
Apples & Applesauce	36	Cranberry Sauce	12+
		Fruit Salad	36
Apricots	36	Grapes	12+
Blackberries	12+	Peaches	36
Blueberries	12+	Pears	36
Cherries, R.S.P.	12+	Pineapple	36
Cherries, sweet	12+	Plums	12+
Cherries, black	12+	Rhubarb	12+
	Vegetables		
Asparagus	36+	Hominy	96+
Beans, Lima	96+	Peas	96
Beans, stringless	36+	Pickles	12+
Beets	48+	Pumpkin and Squash	48+
Brussel Sprouts, Cabbage & Cauliflower	49+	Sauerkraut	12+
		Spinach	36+
Carrots	96+	Sweet Potatoes	48+
Corn	96+	Tomatoes	48+

[1] "Progress in the Tin Plate Industry," *Food Technology*, (Vol. 5, No. 10, Oct. 1951).

Product	Shelf Life (in months)	Product	Shelf Life (in months)
	Other Foods		
Baking powder	24	Margarine	12
Bouillon cubes	12+	Pasta	36
Cereals, ready to eat	9	Pudding Mixes	12
Chocolate, semisweet	24	Rice, white	96+
Condensed and evaporated milk	24	Salad Dressing	9
Dried Beans	96+	Salad Oil	10
Jams, Syrups	36	Shortening	12+
Jell-O	36+	Vegetable Oil	36
		Sugar	96+
		Yeast	12
	Mixes and Packaged Foods		
Cake Mixes	12	Soups	36
Casseroles	18	Onions, minced dried	36
Frosting (in cans)	8	Nuts	9
Hot-Roll Mix	18	Peanut Butter	9
Pancake Mix	6	Soft Drinks	6
Pie Crust Mix	8	Canned meats	48+
Potatoes, Instant	36+	Juice Drinks	6
Herbs & Spices (unopened)	36+		

A Practical Plan for Food-Storage Shelves

by George S. Nelson

Can Storage Shelves

Note: Place cans at rear of shelves. Remove cans from front. Date rear can with crayon.

Capacity: up to 900 cans of assorted sizes.

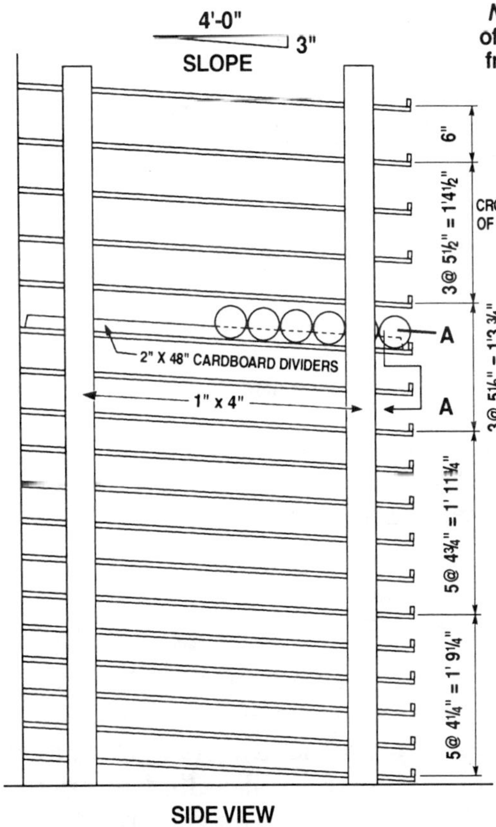

SIDE VIEW

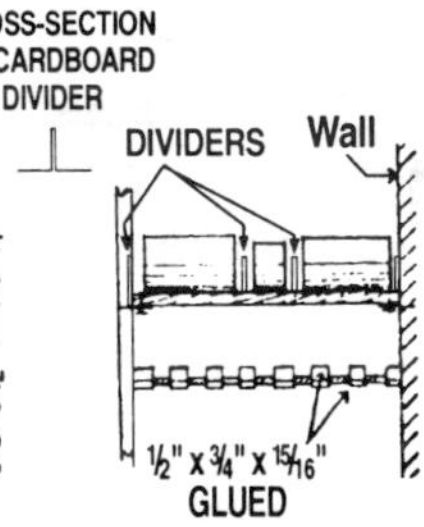

SECT A-A

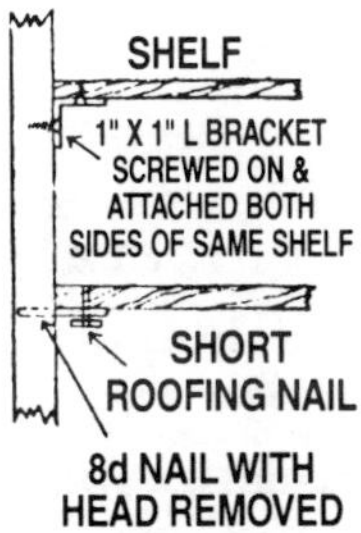

FASTENERS

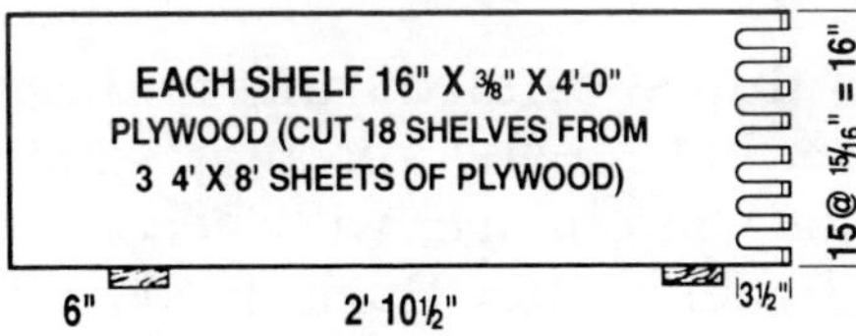

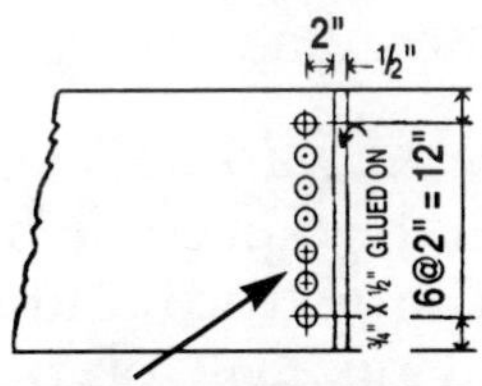

BEFORE CUTTING GROOVES, FIRST MAKE 1" DIAMETER HOLES ABOUT 2" APART FROM CENTER TO CENTER, THEN GLUE ON RETAINING STRIP

SOLVING FOOD STORAGE PROBLEMS

Food has many different enemies which will destroy and spoil the quality and nutritional value of it. Since purchasing in quantity means storing food, let's look at some of these enemies and how we can control them. They are:

1. Temperature
2. Moisture
3. Insects and rodents
4. Time
5. Light

Temperature

Temperature causes a variety of chemical reactions that change the food we eat. As a rule of thumb, for every 20° F. increase, the shelf life of food is decreased by one-half. The best-suited temperature for canned fruits and vegetables is below 70° F. Most basements will maintain this temperature.

At warmer temperatures, canned fruits and vegetables become somewhat mushy and may soften. The texture and color will change, with accompanying loss of nutritional value. Such foods still maintain some nutritional value, however.

Dried foods often become rancid with warm temperatures. This mostly occurs when the natural structure of the food has been broken down, such as cornmeal, flour and polished rice. It is recommended for this reason to store the whole grain and then grind it with a hand or electric grinder. Rancidity may be kept to a minimum by storing the food in sealed containers.

Moisture

Moisture can be a problem with dried foods, grains, and freeze-dried products. Flavor and color is lost when moisture is allowed to condense in the food.

Canned foods pick up moisture when left on a cement floor. Always place wood, cardboard or something between the cement and the food. Moisture also causes soybean meat substitutes to go rancid.

Freeze-dried food should be covered with an airtight seal after it is opened.

The remedy for preventing food from being destroyed by moisture is to store it in sealed, airtight containers. If you have a large amount of food, break it down into smaller airtight containers to store it more safely. You can use large containers from which smaller canisters are filled when needed. Be sure to reseal the container.

Molds, Insects and Rodents

One of the worst effects of moisture is it allows microbial activity. Mold grows when moisture is allowed on dried foods. Moisture and microbial activity destroys, and is the main cause of food spoilage.

Insects and rodents contaminate much more food than they eat. It is important to protect your food from insects and rodents. The best way to protect your food from insects is to seal in air-tight containers. If you are using plastic, caution should be exercised to make sure it is rodent proof. Rodents (rats, not mice) are known to have chewed through plastic buckets.

In grains and flours, the biggest problem is weevil. Wheat should be lower than 10% in moisture content to protect against these creatures. Weevil also thrive on traces of soap remaining in containers which have been washed with soap. Detergent does not have the same attraction to weevil.

Food products may be fumigated with dry ice to protect them from weevil. About 1/2 pound of dry ice per 100 pounds of grain is sufficient. Place the dry ice on the bottom of the container, then pour the wheat or grain over it. Put the lid on loosely, allow the dry ice to send its fumes and cloud up through the grain until it melts. Then seal this lid after two hours. The expanding ice will cause an explosion if you seal the lid right after you pour the wheat over the dry ice.

If you don't have access to dry ice, then you can pour a cup of salt inside your container, or place a small bag of salt inside the container. This will absorb moisture and prevent weevil.

If weevil or larva are present, you can destroy them by heating the grain in an oven at 150° for sufficient time to heat the grain in the center, then sift the weevil out.

Time

Canned foods will remain wholesome and very useable as long as the containers remain intact and are not bulging.

One enemy of long storage upon food and food value in metal cans is the factor that metal attacks the food. Approximately 2%-5% of food value is lost each year, so it's just a matter of time before all food value is lost. Dried fruits will last in a closed dry container as long as it looks and smells good.

Unopened, freeze-dried foods will keep indefinitely. However, the shelf life is reduced to a month or less when opened.

When considering long-time storage, store whole grains instead of flours or cornmeals. The grains maintain their food value, while the flours will break down much faster.

Light

Protein foods such as powdered milk, canned meat and cheese should be stored in a dark, cool place. Protein is said to be destroyed by the presence of light.

Vitamins, especially vitamin C, should be kept in the dark to prevent the loss of nutritional value. This should not be too much of a problem because most vitamins are packed in dark glass containers.

Containers

Most containers are adequate. Glass, plastic and metal are good containers, provided that they are air-tight. Glass should be protected to prevent breakage. All containers should be kept up and off cement floors. Large containers are good to store the basics. Be sure to prevent spillage; if you do spill, clean it up immediately to keep insects out.

FRESH PRODUCE FOR HOME CANNING

We mentioned that one method of purchasing in bulk is to buy produce when it is at its best buy, and then can or freeze it for later use. The most popular fruits and vegetables are usually available all year long. The following chart shows when individual fruits and vegetables are most abundant and therefore the best value. A home garden would provide a great savings on your home canning bill.

Vegetables	**J**	**F**	**M**	**A**	**M**	**J**	**J**	**A**	**S**	**O**	**N**	**D**
Artichokes				X	X							
Asparagus			X	X								
Green Beans					X	X	X					
Broccoli	X	X	X	X								
Brussel Sprouts										X	X	
Cabbage	Good Availability All Year Long											
Carrots	Good Availability All Year Long											
Cauliflower										X	X	X
Celery	Good Availability All Year Long											
Corn					X	X	X	X				
Cucumbers					X	X	X	X				
Egg Plant								X	X			
Green Peas				X	X	X	X					
Green Pepper					X	X	X	X	X			
Potatoes	Good Availability All Year Long											
Squash	Fairly Good Supply All Year Long											
Spinach	Fairly Good Supply All Year Long											
Tomatoes					X	X	X	X				

Fruits	**J**	**M**	**M**	**J**	**S**	**N**
Apples	X	X			X	X
Bananas	G	d A	lab	y A	ear	ng
Melons			X	X		
Oranges	X	X	X			
Peaches				X		
Pears					X	
Strawberries			X			

Freezing Points, Recommended Storage Conditions, and Length of Storage Period of Fruits & Vegetables

Commodity	Freezing Point F.	Ideal Place to Store	Storage Temp.	Storage Humid.	Length of Storage
		Vegetables			
Dry beans and peas	___	Any cool dry place	32-40°	Dry	As long as desired
Late Cabbage	30.4	Pit, trench, or outdoor cellar	Near 32°	Mod. moist	Through late fall and winter
Cauliflower	30.3	Storage cellar	"	"	6 to 8 weeks
Late Celery	31.6	Pit or trench; roots in soil in storage cellar	"	"	Through late fall and winter
Endive	31.9	Roots in soil in storage cellar	"	"	2 to 3 months
Onions	30.6	Any cool, dry place	"	Dry	Through fall and winter
Parsnips	30.4	Where they grew or in storage cellar	"	Moist	"
Peppers	30.7	Unheated basement or room	45-50°	Mod. moist	2 to 3 weeks
Potatoes	30.9	Pit or storage cellar	35-40°	"	Fall-Winter
Pumpkins & Squash	30.5	Home cellar or basement	55°	Mod. Dry	Through fall and winter
Root Crops	___	Pit or storage cellar	Near 32°	Moist	"
Sweet Potatoes	39.7	Home cellar or basement	55-60°	Mod. Dry	"

Commodity	Freezing Point F.	Ideal Place to Store	Storage Temp.	Storage Humid.	Length of Storage
		Fruits			
Apples	29.0	Fruit storage cellar	Near 32°	Mod. moist	Though fall and winter
Grapefruit	29.8	"	"	"	4 to 6 weeks
Grapes	28.1	"	"	"	1 to 2 months
Oranges	30.5	"	"	"	4 to 6 weeks
Pears	29.2	"	"	"	2 to 4 weeks

TIPS ON QUANTITY FOOD BUYING

Does Bulk Buying Save Money?

Before you begin an expensive adventure you must first ask yourself if it is worth it. When purchasing food in quantity you must ask yourself, am I really saving money? Is the time spent cooking, preparing and storing worth it?

The best way that these questions can be answered is to show you how much you can save. These 1999 prices will probably be old by the time this book comes off the press, which will show that by purchasing food you save money because of inflation and increasing prices.

A large three-pound bag of minced onions costs $4.99 in a local (Utah) grocery store while a three-ounce bottle costs $3.68. If you purchased 16 ounces (one pound) in small jars it would cost you $11.04. By purchasing a pound you have saved over 50%.

Honey, in a five-pound bottle, is sold for $6.99. If purchased in small jar size, one pound will cost

$2.49. If you purchased five of those they would cost $12.45—nearly double.

Rolled oats recently cost $.49 per pound sold in bulk. The same product purchased in a two-pound name-brand container would cost you $1.75 per pound, or more than three times as much.

Flour will cost you about $1.89 for a five-pound bag, or $.38 per pound. A 25-pound bag costs $5.99, or $.24 per pound, that's a savings of 14% per pound.

The disadvantages of buying your food supply in quantity are as follows: *space, spoilage, rotation* and the *time spent in preparing meals*.

A person who watches the sales and buys in bulk will save between 20-30% of his food bill. If you wish to count inflation and other factors, you could possibly save between 40-50% on your food bill. This would amount to an annual savings of from $700 and $800 for those who spend between $120 and $140 on their monthly food bill. With large families, the savings would be much greater. It is well worth it!

Here's How It Works

The "buying in quantity" method takes into consideration our grandparents' methods of storing and using foods, and combines it with the modern housewife's supermarket approach. One doesn't need to join any expensive food clubs to save money. Here are some very important steps in bulk buying:

1. *Plan your needs.* Quantity buying means buying a three-month, six-month or even a twelve-month supply at once. You must determine what your family needs and what their likes are. If you cannot afford to buy a year's supply at a time, then try buying for a three-month or six month period.

2. *Watch for sales.* One method is to watch for a good sale on something that you use quite a bit of, such as tuna fish, and then buy several cans of this item at the sale price. You will have saved money this way. You may need to reallocate the family budget to provide for these special sales. Reserve $10 or $20 (or how much you can afford) each pay day to stock up on these items. Don't buy more of these items unless you run low or until there is another sale.

3. The most important thing is to *start with a plan.* When you want to purchase in quantity, you must have a well-planned program. Be sure to choose those items that your family now eats, those items that will store for more than six months, and those items that are nutritionally good.

Try These Different Approaches

The first approach is getting down to the basics. The basic method is used especially when you want the most for your money. You purchase those items that will take longer to prepare, but are less expensive than their already prepared counter-parts. These items include whole wheat, beans, rice and other grains. The basics are often more nutritious than the already prepared foods.

The second approach to consider is purchasing case lots. The retail grocery stores run sales on case lots, and then is a good time to stock up. One may purchase an entire case or enough cans to fill his basic plan. In this method one mainly stores convenient and easy to prepare foods, such as boxed cereal, cake mixes and canned soups.

The third approach to this program, which many housewives now use, at least in part, is to buy fresh produce when in season and can, bottle and freeze these items. In times past, our forefathers grew everything they needed to use. If one had a large garden this would be an ideal situation, because he could become self sufficient if needs be.

Another way is to use a combination of the three above-mentioned approaches. This combination one is probably the best approach to use. You can have the basics as well as the convenience foods. If you have a garden you can save money by growing your own food. One can mix these three approaches in any way he desires.

Since sugar is so expensive, many housewives are now beginning to dry their fruits and vegetables. Dried fruits and vegetables are more nutritious than their counterpart canned and frozen items. One of the best ways to save money on fresh produce is to have a small garden. Our family has saved money on fresh vegetables by having our own garden.

In using these three methods, you can save from 10% to 50% on items that you would normally purchase.

Two Main Ways to Make Quantity Purchases

There are two main ways of buying in quantity or buying in bulk. Instead of purchasing a 3- or 4-ounce package, you purchase a pound or a two-pound package. If you use a lot of a given item, say rice or beans, then you would purchase a 50- or 100- pound bag. The other way to purchase in bulk is to buy more than one at a time, generally a case at a time. If you don't need a full case, then you can just pick up what you need.

The main approach you should take when purchasing in bulk to buy a three-, a six- or a twelve-month supply at once. Don't buy for just a week or a two-week period. You can also purchase in bulk by going in with a group of friends or neighbors. Many wholesale and retail stores give discounts for volume buying. For example, one bag of food may sell for $40, while ten bags will sell for $38 and 25 bags for $36. If you can get in such a group this is a good method to save money.

Other Important Shopping Hints

1. Generally speaking, the smaller the package the higher the cost per weight on the item to be purchased, so you'll save money by buying the larger package.

2. Off brands and store brands are often less expensive than nationally advertised brand names. In fact, many nationally known companies pack these off brands, so check the label.

3. Compare cost with volume. Sometimes weight of an item is not the determining factor. Powdered

milk, for example, sells for $11.49 for a 64-ounce box. Half a gallon (64 oz.) of whole milk costs $2.09. However, you can make 20 quarts of milk from the powdered-milk product, bringing the cost per half gallon to $1.15, a savings of $.94 each half gallon.

4. You'll generally find that paper or plastic bags are less expensive than their counterpart cans and bottles.

5. Specials are important to watch for. Grocery stores run specials to attract customers to their stores. Purchase enough for your supply.

6. You may wish to wait for a certain season to purchase food. Powder eggs are less expensive in the spring but are higher in the winter. Dried crops such as beans and rice are generally lower in the fall than in the summer. When the new crop hits the market there is generally a drop in price to sell the old crop. As the year goes on, the supply decreases and the price goes up, so buy in season.

7. Since you may not be able to use products purchased in bulk in a short time, it is a good idea to break them down into smaller units or packages and use the smaller packages when needed. Repackage the items in durable containers, then refill when needed.

8. Convenience foods are more expensive than do-it-yourself items. If you have the time, do it yourself.

NOTES AND IDEAS ON STORING BULK FOOD

1. Plan to use those items that you purchase, especially those with which you are not familiar. You should use your cookbook to find recipes that will appeal to your family.

2. Basic foods are good to have on hand. These include wheat, beans, rice and other grains. You'll find these are full of vitamins and minerals. Begin to use a little at a time and you'll find that you can't live without them.

3. Some foods don't need to be stored in a cool, dry place. To conserve space or where space is a problem, freeze dried, dehydrated foods, sugar and Jell-O can be placed in areas of your house that are not cool.

4. Use iodized salt to protect pasta products from weevil. Add about one cup of salt to 10-20 pounds of pasta.

5. Canned milk needs to be turned every month or so to prevent it from going lumpy.

6. Baking powder may be kept in the original metal container. Baking soda should be repacked in metal or glass and then sealed.

7. Canned goods need to be turned or moved every six months to prevent settling of products. A good shelf that is self rotating is very helpful.

8. Brown sugar, after being stored for some time, becomes hard; if yours goes hard, place an apple in it for a day.

9. Yeast may be stored in the freezer or refrigerator for longer life. If purchased in a large container, you may want to divide it into smaller packages before storing.

10. Beans, peas and lentils do not require a fumigant, but should be sealed in an airtight container to prevent loss of moisture. If they are quite gassy, try sprouting them before cooking.

11. Fats and oils in metal containers which remain sealed should keep for many years.

12. Dried fruits & raisins should be as dry as possible; then repack them loosely in clean, glass fruit jars or cans which are placed without lids in an oven, heated to 150° for 20 minutes. Cap with sealing caps.

13. Since sugar is expensive, an investment in a fruit dryer would be a good item to cut the cost of home canning.

14. Flour, milk and sugar should be repacked in dry, metal, plastic or glass containers and sealed with airtight lids.

15. If you package in glass, make sure your shelves are protected against falling in case of an earthquake.

16. Be sure to date your food purchases. Use the oldest products first.

17. Never store your canned goods on a cement floor. Place wood, cardboard or plywood on the cement floor before storing canned goods.

KEYS TO SUCCESSFUL FOOD STORAGE

1. ***Quality of the product***. Top quality foods should be obtained whenever possible. Inferior products may be less expensive, but don't keep as well.

2. ***Proper containers***. Containers should be adequate to keep insects and moisture out of the food.

3. ***Storage facilities.*** Areas that are kept cool and dry and permit easy access are the best.

4. ***Temperature.*** Food stores best from 40° F. to 60° F. Foods may be stored at higher temperatures, but shelf life is somewhat less. Consider the temperature of your storage area when you plan your food supply.

5. ***Rotation.*** This minimizes loss of food value and flavor, and prevents the loss of money.

6. Remember these ***points to save money:***

- Plan your storage before you buy
- Rotate your food supply
- Try it before you buy a large volume
- Store only what your family will use

EMERGENCY LIGHT SOURCES

In the event we lost our electric power, we would be left without light. We should have on our storage list some source of light. In summer, the amount of artificial light required would be somewhat less than that required for winter months due to the additional amount of summer daylight. We should also remember that winter storms knock out power more often than other storms.

Figure that you will need about five hours per day of emergency light sources for a single room or location. A two- to four-week supply is recommended.

Candles

Candles have to be used with care, or they pose a significant fire hazard. They should be used in broad-based candle holders or candle lanterns. Aluminum foil can be placed around a candle to reflect its light and shield from breezes which can add to fire danger.

Light-colored candles burn brighter than dark ones. Candles burn more evenly and slowly, and with fewer drippings, if they've been placed in a freezer for several hours before using them.

Tallow candles burn better than wax candles. They're more smoke free, and they burn longer and brighter.

Typical wax candle yields:
- 1/2" diameter x 4" tall = 2 hours and 20 minutes
- 7/8" diameter x 4" tall = 5 hours
- 2" square = 7 hours per inch

Typical tallow candle yields:
- 1/2" diameter x 6" tall = 3 hours
- 1" diameter x 6" tall = 6 hours
- 2" diameter x 9" tall = 48 hours

Emergency candles are also available in canned and bottled form. The canned type utilizes hardened wax. The bottled type uses liquid paraffin which burns without smoke or odor. It burns up to 100 hours and stores indefinitely.

Flashlights

It is recommended that you have several flashlights in good working condition on hand, with one available for each key area of the home. Store extra bulbs for each one. The krypton and halogen types shine most brightly.

Store extra batteries for your flashlight, emergency transistor radio, and other battery-operated items. Alkaline batteries have a shelf life of three to five years if stored in an airtight container in a cool location. Lithium batteries can be stored up to ten years.

Conventional "D" cell dry batteries have the following approximate battery life in hours:

Intermittent use, old batteries:	3-5 hours
Intermittent use, new batteries:	7-8 hours
Continuous use, old batteries:	2-4 hours
Continuous use, new batteries:	5-6 hours

You may wish to purchase one or more self-powered flashlights which are powered by squeezing a lever which causes an internal generator to function. These are tiring to use and don't put out much light, but they're better than nothing in an emergency.

Kerosene Lanterns

A kerosene lantern with a one-inch wick will burn approximately 45 hours per quart of kerosene. This is approximately one ounce of fuel per hour. If a kerosene lantern is burned five hours per day, it will use one-half pint per day, two quarts per week, two gallons per month and 25 gallons in a year. Kerosene lanterns are quite dim, generating about the same amount of light as a sixty-watt light bulb. They're also quite dirty and smoky. Used over a long period of time indoors, they will ruin a good paint job. The glass cover of the lantern needs to be cleaned quite frequently. Kerosene lanterns are the cause of numerous fires and should be used with extra care.

Two-mantle Gas Lanterns

A Coleman-type lantern will burn a gallon of fuel in approximately 40 hours, putting out about the same light as a 200-watt bulb. A gas lantern also generates a considerable amount of heat. If the lantern is burned five hours a day, it will use approximately one quart per day, two gallons per week, eight gallons per month and 100 gallons per year. If a mantle-type lantern is going to be a major source of your emergency lighting, be sure to store a good supply of extra cloth mantles.

Storage and Use Cautions

Do not store Coleman fuel, white gas, kerosene and similar fuels inside the home, near food, or within the reach of children. Do not fill lamps inside the home. Since lanterns exhaust bad air, be sure to use them only where the area is well ventilated. Candles and lanterns all constitute an extremely high fire danger, as do the fuels they consume. Be sure that fuel storage is outside the home. Be aware that explosive gases from butane, propane and natural gas can build up inside an enclosed area. Always have a window open to admit fresh air when using the combustible materials referred to in this chapter.

Be sure to have a good supply of matches on hand. Several automatic cigarette lighters might be handy for lighting candles.

EMERGENCY HEATING

Due to the possibility of fuel-shortage crises, it is wise to store a supply of heat-generating materials to keep our homes warm. Most homes are either electric- or gas-heated, and this supply could be threatened by a variety of circumstances.

It is recommended that you have a two- to four-week supply of emergency heating fuel on hand if you are a city-dweller whose home is normally heated by natural gas or electricity, though a year's supply is a better goal. If you live in a rural setting, you may need to stockpile enough fuel for a full year.

If you are counting on a wood or coal stove, or a fireplace, to heat a portion of your home in an emergency situation, you'll need to develop a realistic estimate of how many logs of firewood or buckets of coal you would use per day in a bitter-cold emergency, and stock accordingly.

Firewood is usually purchased by the cord. A cord is a pile four feet wide, four feet high, and eight feet long, which totals 128 cubic feet. A standard cord of firewood (dry, dense hardwood) weighs about two tons. It provides about the same amount of heat as a ton of coal. Putting a covering over the wood supply such as a tarp or sheeting (plastic or metal) will help to keep the wood pile dry. Supplies for wood or coal stoves would also include flue, pipes, shovel and bucket to remove ashes.

Fireplaces and wood stoves constitute a significant fire danger. You must be sure the chimney has been cleaned. Be careful to keep children away from stoves and fireplaces so that they're safe from burns. Remember that combustible fuel sources consume oxygen and generate carbon monoxide and other deadly fumes.

If coal is used as a major heating source for a home over a long period, you can calculate that two to four tons of coal will be needed, assuming that it will be used sparingly. Coal should be stored in a dark place, and away from circulating air. Moving air tends to break the coal into smaller pieces, which burn faster than larger lumps.

Other Heating Suggestions

Extra blankets are necessary to provide heat at night in emergency situations when not all rooms of the house are being heated. Newspapers may be placed between blankets to insulate heat.

Each year you should check your home's doors, windows and screens to prevent the escape of heat.

Emergency Cooking

You need to plan how you would cook foods in an emergency situation. A typical solution is to count on the use of a Coleman-type two-burner gas stove. You would probably use it about four hours a day in the preparation of three meals. This would utilize five pints of Coleman fuel per day, three gallons per week, twelve gallons per month, and 120 gallons per year. Be sure that all cooking is done in a well-ventilated space, preferably outside the home.

A charcoal grill or small hibachi can also be used, but definitely must be used outside of the home. Three or four charcoal briquettes in a small hibachi should be enough to prepare a simple meal. You would need to store approximately 25 pounds of charcoal briquettes for a week's use. Be sure that you have stored an adequate supply of charcoal starter, too. A propane-gas grill, with several tanks of propane, makes a useable emergency cooking source also.

A small, portable camping stove is a useful item to have on hand. You can also make emergency tin-can stoves out of one-gallon cans or No. 10 cans. You'll need to punch several holes in the sides of the can, both at the top and the bottom, to cause the heat to flow upward. You can set the stove over a small tuna can with a paraffin-wax-soaked roll of cardboard or newspaper. The lighted can will provide a source of emergency cooking heat for such makeshift stoves.

EMERGENCY MEDICAL SUPPLIES

Medical supply needs will vary widely, depending on family sizes and the ages of family members. This list will help you brainstorm your personal needs.

Critical Personal Needs

It is essential that medicines needed by family members for the care of chronic illnesses (heart conditions, diabetes, etc.) be available for an extended emergency period. You will need to consult with your physician and your pharmacist to acquire these medications in sufficient quantities. Be sure that all medicines are properly labeled, and that directions for their use are on hand. They should be stored in a dark, cool, dry location, and a record of their expiration dates be maintained.

Books & Manuals

- First Aid Book
- Health and Nutrition Book

Basic Medicines

- Antibiotics (broad spectrum, both adult and child strengths, consult physician)
- Aspirin tablets (adult, analgesic)
- Tylenol (adult and children, analgesic)
- Pain medication (strong, consult physician)

- Pepto-Bismol, Imodium, Lomotil, or other medicines (for diarrhea)
- Antihistamines (dry nasal secretions)
- Decongestants (to clear stuffed nasal passages)
- Cough medicines
- Expectorants (to expel chest fluids)
- Laxatives (for constipation)
- Benadryl (for congestion, allergic reactions, itching)
- Lotion (for poison ivy, bites and stings)
- Insect repellent
- Neosporin or equivalent (for skin infections, minor wounds)
- Hydrogen Peroxide (for abrasions, minor cuts)
- Sunscreen
- Vaseline lotion (for dry skin)
- Ear drops (for ear infections)
- Eye drops (for eye infections)
- Toothache medicine
- Salt tablets (for heat emergencies)
- Baking soda (for burns, shock)

Burns

- Industrial burn septic (burn spray)
- Antibiotic Ointment

Cuts & Abrasions

- Antiseptic spray
- Merthiolate spray
- Alcohol preps
- Antiseptic swabs

Broken Bones

- Splints, all sizes and kinds
- Rescue blanket
- Stretcher

Poisons

- Universal Antidote

Bleeding

- Ace Bandages (2- and 3-inch widths)
- Adhesive Tape
- Band Aids, Curads
- Gauze pads and rolls
- Cotton balls
- Compress dressings (2x2, 4x4, 6x6 inches)
- Triangular bandages
- Tourniquet

Other Supplies

- Tweezers
- Safety glasses
- Snake bite kits
- Ammonia Inhalants
- Scissors
- Sterile tongue blades
- Hot water bottle
- Ice bag
- Mouth-to-mouth instructions
- Choking instructions
- CPR instructions
- Rubbing alcohol (70% isopropyl)
- Thermometers (oral and rectal)
- Safety pins
- Paper drinking cups
- Small flashlight with extra batteries
- Q-tip cotton applicators
- Extra contact lenses (with solutions)
- Extra eyeglasses
- Extra hearing-aid batteries
- Matches
- Paper and pencil

PAPER GOODS AND SANITATION SUPPLIES

A supply of the items listed below is necessary in the event of a water shortage or famine to conserve your supply of water. These items are disposable, and will probably be used excessively in the early days of an emergency, especially if water is not available for washing purposes. It is recommended that at least a 60-day supply be on hand, for each member of the family, considerably more if possible. These items will be good trading items in the event of a serious and/or long-term emergency.

Paper Goods

- Toilet paper
- Dishes—paper (several sizes)
- Cups—paper (several sizes)
- Paper towels
- Paper grocery sacks
- Facial tissues
- Napkins
- Pre-moistened towelettes
- Table coverings
- Used newspapers

Plastic Goods

- Plastic forks, spoons and knives
- Large leaf bags
- Trash bags
- Used grocery store bags
- 5-gallon polyethylene buckets with lids (for emergency toilets or waste disposal)
- Plastic sink (5 gallon, with hand pump)
- Solar shower (plastic bag of water that the sun can heat)

Sanitation Supplies

- Portable toilet chemicals (available at recreational vehicle supply stores)
- Porta-potties
- Large metal covered garbage cans (to contain wastes)
- Plastic liners (5-6 gallon size, with twist ties)
- Shovel
- Disinfectants (Lysol, etc.)
- Antibacterial soaps
- Ammonia
- Dish detergent
- Laundry detergent (should be useable in cold or hot water)
- Liquid bleach
- Clothes line and clothes pins
- Scouring pads, sponges
- Rags
- Bar soap
- Feminine supplies
- Baby supplies
- Broom and dust pan
- Mopheads
- Carpet sweeper (manual)

Meeting Emergency Toilet Needs

In many disaster situations, water, plumbing, and sewer lines may not be operable. This may occur in situations like earthquakes, floods, freezing weather, etc. If this type of situation occurs, it will become necessary to dispose of human wastes in a sanitary manner to prevent disease and infections.

If water and sewer lines are disrupted, but the toilets remain useable, the toilet bowl can be lined with plastic bags to collect human wastes. If toilets are not available, it may be necessary to improvise indoor toilets using buckets, cans or porta-potties. Outdoor latrines may be constructed, with canvas tarps erected to provide privacy.

CLOTHES

It is advisable to have a year's supply of clothing on hand for all family members. This requires attention to the needs imposed by the various seasons experienced in your locality. In anticipating a long-term emergency situation in which new clothing could not be purchased, emphasis needs to be placed on durable, rugged clothing that could be used in manual-labor situations. Pay attention to good, long-lasting walking shoes, cold-weather items, and layered clothing that can be worn both indoors and outside when limited fuel situations would leave homes partially unheated.

Where children are grown, it is much easier to store clothes. Since children are always growing out of clothing, larger sizes or cloth needs to be stored to provide for their needs.

Making your own childrens' clothes seems to be a good answer to meeting the needs of growing children. You may wish to purchase cloth that is in style and make clothes then purchase more cloth, thus rotating your stock.

Here are some suggestions of items to store:

- cloth: wool
  - cotton
  - corduroy
  - denim
  - perma press
- buttons
- scissors
- zippers
- pins
- shoes
- safety pins
- needles, thread
- shoe laces
- underwear
- socks

GARDENING

Garden tools are a must for every storage supply. One-year's supply of food is geared to provide food for a family until they can provide their own. The following gardening supplies will be needed to prepare to supply subsequent year's food storage. If necessary, the lawn may be dug up to make room for a garden.

Item	Suggested Quantity
• Wheelbarrow	1
• Shovel	2 plus extra handle
• Hoe	2 plus extra handle
• Spade	1 or 2
• Pruning shears	1 or 2
• Rake	2
• Egg cartons	5-10
or seed starters*	40-60
• Garden seeds	15-20 packages
• Canning jars	6 doz. quarts
	12 doz. pints
• Lids and rings	18 doz. lids
• Pressure cooker or canner	1
• Paraffin	1/2 - 1 case

* Egg cartons may be used to start indoor plants, tomatoes and peppers.

Planter pots may be used indoors to supply some fresh vegetables and herbs during winter months.

BASIC SURVIVAL LIBRARY

Horizon Publishers and Distributors is a major producer of books on food storage and personal preparedness. The titles listed here are excellent books, each one providing valuable insights into the areas it discusses. Please write for current price and ordering information.

The ABC's of Home Food Dehydration

Barbara Densley

This excellent guide explains the complete drying process for the novice. It also provides many helpful hints for making the best possible use of a dehydrator. The book includes effective, proven techniques for preserving fruits, herbs, meat, fish, and vegetables. Ninety-two tasty, easy-to-prepare recipes are included, covering fruits, vegetables, meats, croutons, fish, fruit leather, granola, jerky, marinades, herbs, yogurt, deserts, and more. 5 1/2" x 8 1/2" paperback, 128 pages. (#1201) $8.98*

*All prices are subject to change without notice.

Basic Preparedness: A How-to Guide For Preparedness & Self-Reliant Living

The Survival Center, Inc.

This is a comprehensive guide to storage of food, water and medical supplies; sanitation; alternative energy; self-defense and other critical topics. Distributed item.

8½" x 11" paperback, 208 pages. (#6505) $18.98

Beginners Guide to Family Preparedness: Food Storage, Back to Basics and Survival Facts

Rosalie Mason

This book is a thorough, practical guide to help families prepare for unforeseen natural calamities. It explains how to start, maintain, and sustain a food storage program to fit specific family needs. Eight sections address pertinent questions about home food storage, effects of environment on food quality in storage, food dehydration, basic medical supplies, survival mechanics, and survival gardening. This is one of the best books available for those beginning their family storage and readiness program. Fifty delicious recipes show how to prepare and enjoy dehydrated and stored foods. 5½" x 8½" paperback, 160 pages. (#1217) $12.98

Creative Wheat Cookery

Evelyn C. Ethington

Here are over 300 easy tips, tasty recipes and low-cost ideas for using wheat and gluten in the home. The book tells all about wheat: how to sprout it, store it, benefit most from it, and various methods for cooking it. Vegetarians will enjoy the recipes for using gluten as a healthful meat substitute. Practical suggestions are made for using gluten as a meat extender, and the book is loaded with 164 delicious recipes! 5½" x 8½" paperback, 112 pages. (#1242) $7.98

The Desert Shall Blossom— A Comprehensive Guide to Vegetable Gardening in the Mountain West

David E. Whiting

This guidebook is a comprehensive guide to vegetable gardening in the Mountain West. It offers clear and detailed instructions for planning and maintaining a bountiful garden in the Mountain West's arid, alkaline soils. Suggestions on fertilizers, weed and insect control are based on environmental safety and make this a vital, very comprehensive gardening manual. 8½" x 11" paperback, 112 pages. (#1239) $14.98

Doin' Dutch Oven—Inside and Out

Robert L. Ririe

More than one hundred delectable Dutch oven recipes, plus menus and details on using Dutch ovens on the stove at home and in emergencies, as well as at outdoor campsites, make this an ideal guide for both beginners and experienced Dutch oven cooks. It's filled with instructions and tips that help you master the basics of using a Dutch oven effectively. Recipes include appetizers, dips, breakfast specials, main dishes, casseroles, soups, and side dishes. Every Dutch oven cook can enjoy this book! 5½" x 8½" paperback, 128 pages. (#1218) $9.98

Dutch Oven Secrets

Lynn Hopkins

Written by a veteran of several world championship Dutch oven competitions, this book explains every essential aspect of this unique and delicious style of cooking. He shows how to select a Dutch oven, provides a list of helpful accessories, and teaches the right way of seasoning, using, cleaning and storing the Dutch oven, and even how to prepare the fire for the best results. Also included are 84 of the author's favorite recipes. This is an excellent guide to Dutch Oven cooking! 5½" x 8½" paperback, 112 pages. (#1214) $9.98

Dry Peas and Lentils— Cooking with the "Protein Twins"

Betty Janson

The legume family (peas, lentils, beans, etc.) is one of nature's great gifts to man. This book is a valuable compilation of 199 delicious, tested recipes using dry peas, split peas, lentils, chick peas and other legumes year-around. It's packed with economical, nutritious, and taste-tested recipes for soups, main dishes, appetizers, snacks, salads, dressings, even desserts and breads, all based on the versatile pea and lentil. This is an ideal book for every cook and for every home preparedness library. 6" x 9" paperback, 218 pages. (#1238) $14.98

Emergency Preparedness: A Practical, Common-sense Guide

Evan Gabrielsen

Written by a newspaper columnist, every chapter of this book has appeared and has been well-received as feature articles in various mountain-west newspapers. This book is filled with dozens of tips that inspire action. It's level-headed thinking, with valuable new insights that families and communities can use to real advantage to dilute the danger and suffering of the potential emergencies that face us all.

6" x 9" paperback, 160 pages. (#1252) $11.98

Emergency Survival Packs— How to Prepare 72-Hour and 14-Day Family Evacuation Kits

Blair D. Jaynes

The need to suddenly evacuate, or to function without being able to buy food or other supplies for several days, is experienced by hundreds of thousands of people every year. Will you be ready when your emergency suddenly occurs? Written by a former military survival expert, this book is a well-illustrated guide to preparing an evacuation pack that could literally be a life-saver in the event of an earthquake, fire, flood, or other disaster. Included in this important book are lists of essential materials various-sized families would need, as well as details on emergency necessities listed by age group. This book can help you make lemonade out of a lemon if you're prepared for that sudden emergency!
5½" x 8½" paperback, 32 pages. (#1230) $5.98

40 Keys To Family Emergency Readiness— C-60 Audiocassette Tape

Duane S. Crowther

This audiotape offers an extensive checklist to help the listener prepare ahead for any emergency. Its important keys include a family evacuation kit, a personal preparedness plan, emergency lighting, cooking and medicine, emergency cash and transportation, a neighborhood watch system, first aid training, money management fundamentals and survival skills.
C-60 audio tape with written outline. (#1810) $13.98

Fun With Fruit Preservation—Leather, Drying & Other Methods

Dora D. Flack

Complete with 110 tasty recipes, this book gives details on making fruit leather (which alone makes this book extremely popular); drying and dehydrating a variety of foods; canning and bottling, freezing, making jams, jellies, juices, and syrups; and storing foods in basements, cellars, and pits. This is an excellent guide for all who wish to master the home canning and bottling processes. 5½" x 8½" paperback, 112 pages. (#1204) $7.98

Home Food Dehydrating—Economical "Do-It-Yourself" Methods

Jay & Shirley Bills

This book is full of instructions on dehydrating meats, fish, fruits, vegetables, and herbs. It shows how to make fruit leather, explains proper use of preservatives, and describes ancient and modern dehydrating methods. There are 163 tested, simple recipes for using all types of dehydrated foods are included, and many people recognize it as one of the best dehydrated-foods recipe books available. It's excellent for all who have purchased large quantities of dehydrated foods from commercial suppliers. 5½" x 8½" paperback, 160 pages. (#1205) $10.98

Home Garden Hints

Alan K. Briscoe

This simple but useful booklet provides guidance on planning and planting your garden; on insect, disease and weed control; on transplants; on fertilizing, composting and mulching; on companion planting and hybrids; and useful information on raising specific vegetables. This is an invaluable guide for the home gardener, presenting loads of information in a concise, useful format! 5½" x 8½" paperback, 48 pages. (#1206) $5.98

How To Be Safe in an Unsafe World

Donald D. Gilbert

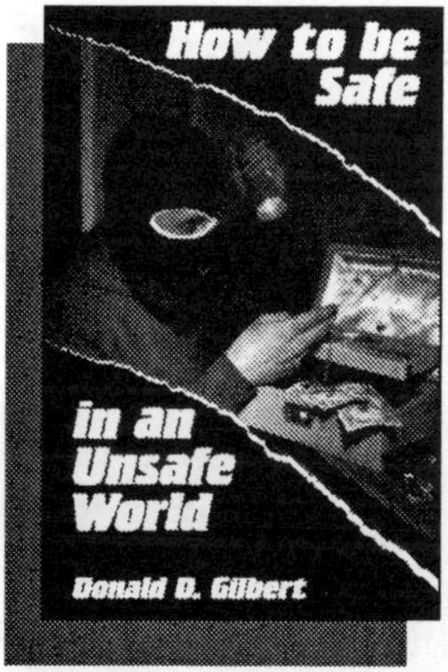

The author, a security consultant, offers valuable instructions on securing entrance doors, windows, and locks in the home, selecting and strategically installing lights and alarms, and setting up security fencing. Advice is also given on training guard dogs and organizing neighborhood watches to ensure maximum home security. Important information on special security considerations for farms and small businesses is also included in this helpful guide. Its information is valuable for both city and country dwellers—everyone can benefit from its useful recommendations! 6" x 9" hardbound, 112 pages. (#1251) $10.98

Just Add Water—
How to Use Dehydrated Foods and TVP

Barbara G. Salsbury

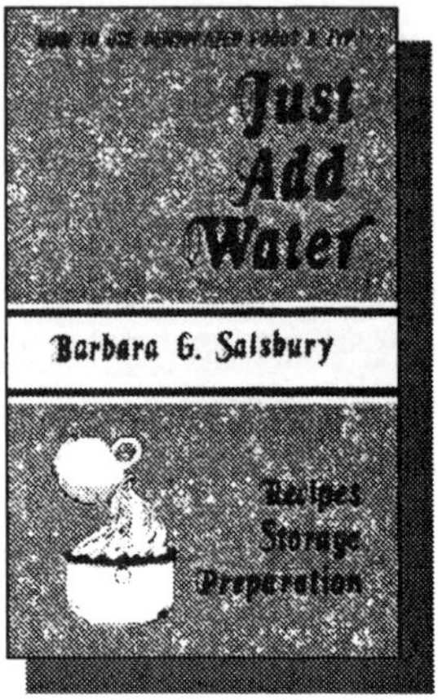

Here's the perfect guide to help the reader prepare dehydrated foods and use TVP (textured vegetable protein–made from soybeans). The 158 recipes are simple and delicious. Valuable hints on packaging and shelf life are included, along with storage plans that make the best use of available space. Suppliers of dehydrated foods have long considered this book as the best recipe book on storage preparation to recommend to consumers. $5^1/_2$" x $8^1/_2$" paperback, 96 pages. (#1207) $8.98

Let's Cook Dutch—
A Complete Guide
for the Dutch Oven Chef

Robert L. Ririe

A Dutch oven is an essential part of emergency preparedness. This basic but comprehensive book is so good that a major Dutch-oven manufacturer supplies it with every oven it sells! It shows how to prepare, use, and maintain the oven, and also teaches essential fire-preparation techniques. Mouth-watering recipes for meats, vegetables, breads, and desserts are included. This book also includes suggestions for Dutch oven cooking in the kitchen, backyard, or in camp for large groups. Cooking with Dutch ovens in unforeseen emergencies is covered also. This is the best basic Dutch oven book available! $5^1/_2$" x $8^1/_2$" paperback, 104 pages. (#1220) $8.98

New Concepts In Dehydrated Food Cookery—Hundreds of New Ideas and Tested Recipes for Enjoying Home Dehydrated Foods

Barbara Densley

More than 350 recipes provide tasty and interesting new ways of using a variety of nutritious dehydrated foods, including meats and fruits. This guidebook thoroughly explains the principles behind pre-treatment of foods being dehydrated and teaches the best drying techniques. It also shows how to use solar and microwave dryers plus the most efficient new methods of food preparation and storage. This book is unique in its combining reconstituted dehydrated foods with fresh foods to enhance taste, nutritional yield, ease of preparation and variety. It's an excellent source for talented, creative culinary artists!
5½" x 8½" paperback, 192 pages. (#1221) $13.98

Prepare Today—Survive Tomorrow

Joe Harold

Written by an expert on survival methods for use in unforeseen natural or political crises, this book provides essential information designed to help individuals and families prepare for and successfully cope with potentially catastrophic events. The author's information and suggestions go far beyond what is generally available. He includes practical advice on raising and storing basic foods, fallout shelter construction, weapons for defense, and how to prepare for potential attacks on both a neighborhood and individual level. This book also suggests valuable personal and family defense philosophies. 6" x 9" paperback, 192 pages. (#1250) $14.98

Project: Readiness—Guide to Family Emergency Preparedness

Louise E. Nelson

This book is written to help families to be prepared for any natural or man-made disaster that might occur. It is organized in four major sections: Emergency and Disaster Planning, Food Storage, Storing Non-Food Items, and Other Aspects of Emergency Preparation. In addition to providing detailed information concerning how to prepare, survive, and recover from many types of disaster situations, it is undoubtedly among the most comprehensive collections of information on food storage in print. It considers the what, where, and how, with details on many items such as emergency water supplies, home food preservation techniques, detecting spoilage, sprouting, pest prevention, fruit leather, and 87 basic recipes. The final sections treat baby and child care, emergency recreation and much, much more. 6" x 9" paperback, 270 pages. (#1243) $18.98

Skills For Survival—How Families Can Prepare

Esther Dickey

Written by the same author who wrote the popular *Passport to Survival*, this book covers the basics for managing for personal and small-group subsistence in times of danger, scarcity and hardship. It explains the "how to" of making emergency clothing, distilling safe water, and storing food, with 166 tasty recipes included for using stored foods. It's also a valuable self-sufficiency gardening guide, with suggestions for extending harvests, overcoming drought conditions, doctoring plants, and more, all of which are of value even in times of plenty. 6" x 9" paperback, 224 pages. (#1219) $17.98

Soybean Granule Recipes

Alan K. Briscoe

Looking for a way to stretch your food budget dollars and provide a healthy alternative to meat products? Soybean granules (TVP) are very rich in protein, low in fat and calories and contain no cholesterol. They can easily be used in place of meat or added to meat to give you savings per pound. This booklet contains information for substitutions and 35 flavorful recipes. $5^1/_2$" x $8^1/_2$" paperback, 32 pages. (#1209) $4.98

Sprouting for All Seasons— How and What to Sprout, Including Delicious, Easy-to-Prepare Recipes

Bertha B Larimore

The author, a prize-winning cook, presents 220 tested recipes for using delicious and nutritious sprouts effectively in all kinds of dishes, from breads to desserts. She thoroughly covers six proven sprouting techniques and provides tips for sprouting wheat, sunflowers, and many other seeds. $5^1/_2$" x $8^1/_2$" paperback, 140 pages. (#1203) $11.98

Tasty Imitations—A Practical Guide to TVP and Meat Substitutes

Barbara G. Salsbury

Learn to use food storage without sacrificing nutrition or taste quality in your meals! This book offers 138 great kitchen-tested recipes for beef, bacon, chicken, ham, and sausage-flavored TVP (textured vegetable protein) or TSP (textured soybean protein), as well as regular grocery items. A section on the use of dried beans and other legumes makes this one of the most practical and essential guides to the use of long-term storage. And as a recipe book for TVP, it's unequaled in its field! $5^1/_2$" x $8^1/_2$" paperback, 80 pages. (#1244) $7.98

You Can Stay Alive—Wilderness Living and Emergency Survival

Larry J. Wells & Roger D. Giles

This practical book focuses on what the inexperienced person must know to survive in outdoor emergencies. The authors explain how to stay alive, healthy, and mentally stable in a crisis. They carefully address the essentials of basic first aid, outdoor living, overland travel, building sturdy shelters and making fires, locating palatable water, finding and preparing food, and even how to make contact with rescue teams. This book belongs in the vehicle glovebox of every camper, traveler and outdoorsman. $5^1/_2$" x $8^1/_2$" paperback, 128 pages. (#1228) $9.98

Earthquake Safety

Ellen Jackson

Written so that even children can understand, this book tells what to do in the event of an earthquake at home, in school or outdoors. Includes checklist for supplies. 8½" x 11" saddlestitched. (#1331) $5.98

Household Safety

Ellen Jackson

Step-by-step help on first-aid treatment, home safety and security. Includes what to do in case of fire, drowning or poisoning, and how to treat for electrical shock. 8½" x 11" saddlestitched. (#1332) $5.98

Stranger Danger

Ellen Jackson

Information and quizzes to help children recognize and avoid the improper approaches a stranger may make on the street or elsewhere. 8½" x 11" saddlestitched. (#1333) $5.98

INDEX